AF564875

Essentials of Agricultural Sciences

About the Editors

Ankit Kumar Maurya is currently pursuing Ph.D. in Agricultural Economics from Chandra Shekhar Azad University of Agriculture and Technology, Kanpur. He has accomplished B. Sc. (Ag.) from Udai Pratap Autonomous College, Varanasi (U.P.) and M.Sc.(Ag.) in Agricultural Economics from Banda University of Agriculture and Technology, Banda (U.P.). Mr. Ankit has got many awards and also has various publications in reputed national and international journals.

Dr. Rahul Kumar Rai did his B.Sc. (Ag.) from SDJPG, College, Chandesar, Azamgarh (U.P.), M.Sc. (Ag. Agricultural Economics from Udai Pratap Autonomous College, Varanasi (U.P.) during M.Sc. he got first position. And did Ph.D in (Ag.) Agricultural Economics from Chandra Shekhar Azad University of Agriculture and Technology, Kanpur (U.P.). He has more than 14 years teaching, research, extention and administratve experience as Project Associate, Senior Research Fellow, Research Associate, Post-Doctoral Fellow, Scientific Officer (Agricultural Economics) and Assistant Professor and Ex. Incharge Head at IIT, Kanpur (U.P.), JNKVV, Jabalpur (M.P.), VPKAS, Almora (U.P.), KAB-II, New Delhi, UPCAR, Lucknow (U.P.) and Department of Agricultural Economics, CoA, BUAT, Banda (U.P.) Furtheremore, Dr. Rai has worked as Chairman, Member and Member Secoretory of various committees and also received many honours/awards as well as published several research papers, books, book chapters, technical bulletins, lead papers articles and abstracts.

Dr. Yash Gautam is an Assistant Professor in Department of Agricultural Economics, College of Agriculture, Banda University of Agriculture and Technology, Banda. He has done his graduation, post-graduation from Banaras Hindu University. His doctorate was also from Banaras Hindu University where he worked on Economic and Environmental benefits of Solar irrigation pump in Rajasthan. He has teaching experience of five years. Dr. Yash has several research papers in highly rated national and international journals which are indexed in Scopus and Web of science along with one book, e practical manuals, five book chapters and popular articles. He has Young Scientist Award and Young Environmental Award to his credit by various organizations. He has also presented papers in several national and international conferences for which he has received Best Paper Presentation awards. Currently, he is handling important projects funded by UPCAR, NABARD and RKVY as Principal Investigator and Co- Principal Investigator. He has also contributed as Co- Project Director in a project funded by ICSSR, New Delhi.

Himanshu Panday did his B.Sc. (Ag.) from CBGPG, College, B.K.T., Lucknow (U.P.), M.Sc. (Ag.) in Agricultural Economics from Banda University of Agriculture and Technology Banda, (U.P). Presently he is working as Young Professional–II at ICAR-Indian Institute of Sugarcane Research, Lucknow (U.P). Mr. Himanshu has also published 02 International and 04 national research papers, 03 review papers, 06 book chapters, more than 30 articles, and 17 abstracts.

Dr. Shravan Kumar Maurya is Assistant Professor (Agronomy) at ITM University, Gwalior (M.P.). He has completed Ph.D. He has accomplished B.Sc. (Ag.) from Mahatma Gandhi Kashi Vidyapith, Varanasi (U.P.) and M.Sc. (Ag.) Agronomy from Banda University of Agriculture and Technology, Banda (U.P.) Mr. Shravan has got Best Master's Thesis award and Young Agronomist Award and published more than 20 research papers, 2 books, several book chapters and articles.

Essentials of Agricultural Sciences

Ankit Kumar Maurya
Department of Agricultural Economics and Statistics
College of Agriculture
Chandra Shekhar Azad University of Agriculture and Technology
Kanpur-208002, Uttar Pradesh

Rahul Kumar Rai
Department of Agricultural Economics
College of Agriculture
Banda University of Agriculture and Technology
Banda-210001, Uttar Pradesh

Yash Gautam
Department of Agricultural Economics
College of Agriculture
Banda University of Agriculture and Technology
Banda-210001, Uttar Pradesh

Himanshu Panday
ICAR-Indian Institute of Sugarcane Research
Lucknow-226002, Uttar Pradesh

Shravan Kumar Maurya
Department of Agronomy
ITM University, Gwalior (M.P.)

1168, Sector 13, Urban Estate
Karnal-132 001, Haryana
Tel: 91-84470 75807, 18440 41168
Email: contentvibesppa@gmail.com
www.contentvibes.in

Print ISBN: 978-81-97682-56-8

ebooks ISBN: 978-81-97682-53-7

Preface

Essentials of agricultural practices hold paramount importance both presently and in the future due to their pivotal role in sustaining global food security, environmental conservation and economic development.

In the present, agricultural practices are indispensable for meeting the nutritional needs of a growing global population, expected to reach nearly 10 billion by 2050. With arable land becoming scarcer and climate change threatening crop yields, optimizing agricultural practices is essential for maximizing productivity while minimizing environmental degradation. Sustainable practices such as crop rotation, integrated pest management, and soil conservation not only ensure current food production but also preserve resources for future generations. Moreover, agricultural practices play a crucial role in rural economies, particularly in developing countries where agriculture is a significant source of employment and income. By adopting modern techniques such as precision agriculture and agroforestry, farmers can enhance efficiency, reduce production costs, and increase profitability, thereby contributing to poverty alleviation and economic growth.

Looking to the future, the importance of agricultural practices will only intensify as global challenges such as climate change, water scarcity, and land degradation continue to escalate. Sustainable agriculture will be key to mitigating these challenges by promoting resilience, adaptability, and biodiversity conservation etc. Emerging technologies like precision farming genetic engineering, and digital agriculture offer promising solutions for enhancing productivity while minimizing environmental impact. Furthermore as consumers increasingly prioritize healthy and sustainably produced food there will be a growing demand for agricultural practices that prioritize soil health biodiversity, and animal welfare. Agro-ecological approaches organic farming and regenerative agriculture are gaining momentum as viable alternatives to conventional farming methods, emphasizing holistic management practices that enhance ecosystem services and long-term sustainability.

The essentials of agricultural practices are indispensable for ensuring food security, environmental sustainability and economic prosperity on a global scale. Embracing innovation, promoting sustainability and fostering collaboration among stakeholders will be crucial in addressing the multifaceted challenges facing agriculture and securing a prosperous future for generations to come.

Editors

Contents

1

Enhancing the Quality of Human Resource in Agriculture through Skill Development

Latika Sharma [1] ***and S.K. Sharma*** [2]

[1]*Department of Agricultural Economics and Farm Management*
CoA, MPUAT, Udaipur, Rajasthan
[2]*Human Resource Mangement Unit, ICAR, New Delhi*

Abstract

Enhancing the quality of human resources in agriculture through skill development is paramount for sustainable agricultural development. Skilled workers in agriculture can significantly improve productivity, efficiency, and innovation within the sector. Firstly, skill development fosters technological adoption and innovation in farming practices, leading to higher yields and better resource management. This is crucial in addressing challenges such as food security and climate change resilience. Secondly, trained agricultural workers are better equipped to adopt modern techniques and technologies, such as precision farming and organic farming methods, which can improve the quality and safety of agricultural produce. Furthermore, investing in skill development enhances the livelihoods of agricultural workers by enabling them to access higher-value markets and diversify their income sources. Overall, enhancing the quality of human resources in agriculture through skill development not only improves the sector's productivity and efficiency but also contributes to rural development, poverty reduction, and sustainable food production. It is an essential step towards building resilient and prosperous agricultural systems for the future.

Keywords: *Enhancing, Innovation, Skill Development, Resource Management, etc.*

Introduction

India's policy in agriculture sector began in 1950 with a conscious effort to meet out the food needs of Indian population in view of famines but later on shifted to develop the agriculture education and research system for enhancing the productivity of staple crops and develop competent human resource to build skilled man power to strengthen agriculture and solve the problem of food & nutritional security at the nation level (1960,s to 1990,s) to the solution of global problems in agriculture through global solutions (2000,s onward) by forging the regional and global international corporation at different levels.

The closing years of the second decade of 21st century brought one of the most remarkable transitions in the international cooperation for strengthening the agriculture research and education in the history of agriculture in India. In most of the world the transition from a resource-based to a science-based system of agriculture is occurring within a single century.

Science, Technology and Skill Development in 21st Century

In the twenty first century, science and technology are viewed as the drivers of Indian economic growth; and agricultural R&D is expected to play a significant role in the process. Agriculture research and development is being affected by research and development at international level.

Nurturing Growth: The Power of Agri-Entrepreneurship in Modern Agriculture

In recent years, the world has witnessed a dynamic shift in the field of agriculture. Traditional farming practices are evolving, giving rise to a new breed of visionaries known as Agri-entrepreneurs. These individuals are combining innovative technologies, sustainable practices, and business acumen to revolutionize the way we cultivate, distribute, and consume food. Agri-entrepreneurship is not just about farming; it's a multifaceted approach that addresses global food security, environmental sustainability, and economic growth. This article explores the essence of Agri-entrepreneurship and its significance in modern agriculture. Entrepreneurship is one of the key drivers for economic development. During an economic crisis, the importance of entrepreneurship development increases. Entrepreneurship has been linked to improved growth, increased wealth and quality of life. In developing countries like India, planning and implementation for development of entrepreneurial programs are essential for raising the living standard of the vast majority of the backward regions because of their over-dependence on agriculture for employment.

Agriculture is considered as the main economic activity which adds to the overall wealth of the country. In the past, agriculture was seen as a low-tech industry dominated by numerous small family firms, which are mostly focused on doing things better rather than doing new things. However, over the last two decades, this situation has changed dramatically due to economic liberalization and a fast-changing society. Agricultural companies have to adapt to the erratic demands of the market, varying consumer habits, stringent environmental regulations, new requirements for product quality, food safety sustainability, and so on. These changes have opened the way for new entrants, innovation, and portfolio entrepreneurship. Farmers, researchers, agricultural business and governments have recognized this and emphasized for a more entrepreneurial environment in the farming business. Agricultural entrepreneurship has a significant impact on business growth and survival. Therefore, it calls both small scale and large-scale farmers to practice entrepreneurial agriculture.

Employment in Agriculture

To capitalize on its demographic dividend, India must create well-paying, high productivity jobs. Of India's total workforce of about 52 crores, agriculture employed nearly 49 per cent while contributing only 15 per cent of the GVA.

Comprehensive modernization of agriculture and allied sectors are needed urgently. In contrast, only about 29 per cent of China's workforce was employed in agriculture. Industry and services accounted for 13.7 and 37.5 per cent of employment while making up for 23 per cent and 62 per cent of GVA, respectively. A significant number of workers, currently employed in agriculture, will move out in search of jobs in other areas. This will be in addition to the new entrants to the labour force as a result of population growth. By some estimates, the Indian economy will need to generate nearly 70 lakh jobs annually to absorb the net addition to the workforce. Taking into account the shift of labour force from low productivity employment, 80-90 lakhs new jobs will be needed in the coming years. Micro and small-sized firms as well as informal sector firms dominate the employment landscape in India.

As per the National Sample Survey (NSS) 73rd round, for the period 2015-16, there were 6.34 crore unincorporated non-agricultural micro, small and medium enterprises (MSMEs) in the country engaged in different economic activities providing employment to 11.10 crore workers. A large majority of these firms are in the unorganized sector. By some estimates, India's informal sector employs approximately 85 per cent of all workers. India also exhibits a low and declining female labour force participation rate. The female labour force participation rate in India was 23.7 per cent in 2011-12 compared to 61 per cent in China, 56 per cent in the United States.

Enhancing the Quality of Human Resource in Agriculture

According to the 2022 revision of the World Population Prospects the population stood at 1,407,563,842. India has more than 50% of its population below the age of 25 and more than 65% below the age of 35. Enhancing the quality of human resource is an essential pre-requisite for implementing and upgrading research programs, developing technologies, evolving institutional arrangements to face challenges and harness opportunities. Vertical integration of agricultural education is the key to improve the quality of human resources. The lack of qualified manpower in adequate numbers in the frontier areas of agricultural science and technology is one of the major constraints to deliver at grassroots levels for achieving farm prosperity and in taking forward the desired growth rate in agriculture sector. There is a direct relationship between investment in education and poverty elevation as it helps in sustenance of agricultural productivity and profitability. SAUs need to play a greater and more proactive role in attracting best talent by providing enabling environment and facilities at the colleges for the all-round development, growth and employment of students. States should provide more funds to the SAUs to modernize and strengthen the infrastructural facilities to better equip themselves to meet the newer challenges in the sector. To achieve average annual growth above 4%, allied sectors have to be taken on board as we would require not only food but the balanced diet with adequate nutrition. For increasing production and productivity on sustainable basis, the new technologies developed through concerted R&D efforts are going to hold the key for success. Some bigger initiatives with mega investments would be required to produce the competent human resource, not only for the research and teaching purposes but also to effectively deliver the technologies at the grassroots level to give momentum to get the optimum production.

Need of Skill Development

For harnessing the demographic advantage that it enjoys, India needs to build the capacity and infrastructure for skilling/reskilling/up-skilling existing and new entrants to the labour force. The goals to be met until 2022-23 are as follows

- Increase the proportion of formally skilled labour from the current 5.4 per cent of India's workforce to at least 15 per cent.
- Ensure inclusivity and reduce divisions based on gender, location, organized/unorganized,
- India's skill development infrastructure should be brought on par with global standards by:

1. Developing internationally compliant National Occupation Standards (NOS) and the Qualification Packs (QP) that define a job role.
2. Making all training compliant with the National Skills Qualification Framework (NSQF).
3. Anticipating future skill needs to adapt skill development courses.

- Skill development should be made an integral part of the secondary school curriculum.

Present Status

According to the National Policy for Skill Development and Entrepreneurship, more than 54 per cent of India's population is below 25 years of age and 62 per cent of India's population is aged between 15 and 59 years. This demographic dividend is expected to last for the next 25 years. With most of the developed world experiencing an aging population, India has the opportunity to supply skilled labour globally and become the world's skill capital. However, the demographic advantage might turn into a demographic disaster if the skills set of both new entrants and the existing workforce do not match industry requirements. Recognizing the challenge, the Government of India has launched many initiatives to equip fresh entrants with relevant skills and to upgrade the skills of the existing workforce. A dedicated Ministry of Skill Development and Entrepreneurship (MSDE) was set up in 2014 to implement the National Skill Development Mission, which envisions skilling at scale with speed and standards. On July 15, 2015, on the first ever World Youth Skills Day, the Honourable Prime Minister launched the Skill India scheme. To improve the relevance and quality of courses offered by Industrial Training Institutes (ITIs), polytechnics and private training providers, Sector Skill Councils (SSCs) have been involved in curriculum up-gradation/preparation, and in the assessment and certification process. Courses are being aligned to the National Skills Qualifications Framework (NSQF). Recognition of Prior Learning (RPL) has been introduced to ensure certification of and bridge training for the existing work force.

The National Skill Development Policy estimates that only 5.4 per cent of the workforce in India has undergone formal skill training as compared to 68 per cent in the UK, 75 per cent in Germany and 96 per cent in South Korea. The India Skill Report 2018 states that only 47 per cent of those coming out of higher educational institutions are employable.8 Given that 83 per cent of the workforce is engaged in the unorganized sector with limited training facilities, upgrading of skills, both in manufacturing and services sectors remains a challenge

Importance of Agriculture Sector and its Role in Employment

For India, 2022 was special. It marked the 75th year of India's Independence. India becomes the world's fifth largest economy, measured in Dollars with nominal Gross Domestic Product (GDP) of India being around US$3.5 trillion. As per third advance estimates by Ministry of Agriculture and Farmers Welfare India had a record food grain production of about 330.53 million tons in 2022-23 with record all time high production of 135.54 Million ton rice and 112.74 million ton wheat. The fundamentals of Indian economy are sound as it enters its Amrit Kaal, the 25 year journey towards its centenary celebrations as a modern, Independent nation.

Population is rising in India; it is currently 17.63% (1.375 billion) of the total world population. This is predicted to increase by 23.6% in 2020 to 1.7 billion in 2050. India will surpass China soon after 2022. Globally, food demand is projected to increase by 100 - 110% between 2005-2050. In India, if there is no change in policy the demand for food grains including pulses is expected to rise by 49.6% in the next three-decades (Chand, 2012). Similar demand for other agricultural commodities is projected to increase by 28.1% in cereals, 96.4% in pulses, 400% in vegetable edible oils, 90.8% in sugar, 123,4% in vegetables, 78.6% in fruits, 132.5% in milk, 110.4% in meat, 77% in eggs and 90.1% in fish (Singh, 2019). However, the projected increase in production is estimated to be smaller for some commodities (e.g., 107% for vegetables, 88% for dairy products, 162% for rice and 179% for beef) than that of consumption (Hamshere et.al., 2014). As a result, imports are projected to increase. Besides rise in population, a four-fold growth of Indian middle class in the last decade, combined with rapid urbanization and higher GDP are creating a higher demand for food. Emphasis is needed to be given on attracting more and more skilled youth in agriculture to make it more vibrant and efficient to meet the demand of future for food and nutritional security. Besides teaching, research and extension, agricultural entrepreneurship is an emerging field. It involves analysing and understanding the strategies of agricultural entrepreneurs, particularly in response to the institutional changes and economic and technological disruptions to which the agricultural industry is subject.

Agriculture entrepreneurship is the present day need to make Indian farms as commercial production enterprise and it is possible only by effective management of agriculture component; soil, seed, water, market requirements. Agriculture and domestic businesses provide about 50% of employment in half of all jobs in developing countries (World Bank, 2012) but do not produce sufficient income to raise people out of poverty. Thus, entrepreneurial actions associated with agriculture generate a solution for growing household incomes. The right business decision-making ability and commercial knowledge, financial credit back up, supported with government regulatory and policy measures would make possible success of the growing needs of agricultural entrepreneurship in India. Agriculture entrepreneurs are innovators, determined, creative leader, always looking for opportunities to improve and expand their business at any level and they like to take calculated risks, and assumes responsibility for both profits and losses. They are passionate about growing his business and constantly looking for new opportunities.

Agripreneurship

Entrepreneurship is the capability to develop ideas and attain success with them. Innovation, ability to accept change and risk and the organization of resources are some of the factors involved in creating a sustainable enterprise. Entrepreneurship is a feasible approach for upward mobility, as a 1% increase in entrepreneurial activities decreases the poverty rate by 2% (Singh, 2014). An entrepreneur as an individual who controls a business with the purpose of growing the business along with leadership and managerial skills necessary for achieving those goals.

Building Entrepreneurial Human Resource

Every individual has the capacity for successful entrepreneurship and agribusiness in India. Indian policies related to youth entrepreneurship have evolved rapidly over the past decade. It is a recognition that the youth force, particularly in agriculture and allied sectors, is key to the effective inclusiveness and engagement of youth and women in improving the livelihood and long-term transformation of the agriculture sector. Further, given the recent policy reforms and the associated challenges, studying opportunities for youth in the agriculture sector becomes paramount to guiding the policy and program implementation process from the youth entrepreneurial perspective. Indian policymakers operating in the agricultural and rural development sectors recognize youth entrepreneurship as a critical driver for transforming these sectors. Policy and program interventions at the national level reflect this recognition.

Recent economic growth in the last two decades in India has also brought the needed preconditions for youth entrepreneurship. Yet the challenges for entering business opportunities for youth in agriculture remain. There are several structural constraints related to access to technology, finance, institutional support, market access, and business mentorship. These challenges are accentuated further by the needed skills and experience relevant for initiating and running businesses, which remain a significant challenge for the youth in rural India.

Emphasizing Capacity Building for Youth Entrepreneurship

Youth entrepreneurship is considered as a strategic approach to integrating rural youth into economic opportunities. Innovations in the use of information and communication technologies (ICT) further enhance such approaches enabling youth to start their businesses. In addition, access to ICT by rural youth can help in adopting technological innovations in agriculture and agribusinesses and in helping to increase the operational efficiency with which they engage in entrepreneurial activities through better access to digital platforms, reducing transaction costs, and increasing market linkages (USAID, 2019).

Youth engagement in agriculture and agribusiness opportunities needs to be context-specific and should recognize several pathways to their prosperity (SFI Advisors, 2019). First, the commercialization of the smallholder pathway quickly helps the rural youth to see agriculture as a profitable business. Linking them with the existing value chains and through the producer-based organizations will further strengthen their market access and information base, including the reach of the extension and advisory services. Second, youth could engage in providing extension services as private extension agents. Third, they could serve as agribusiness entrepreneurs in the input supply operations and in product aggregation and linking farmers to the markets. Finally, rural youth could also take up business opportunities in the processing sector through agribusiness incubators. In all these pathways, they will need the support of the broad policy environment, institutional backup such as a multi stakeholder agribusiness ecosystem, and building of their individual skills that connect a large number of rural youths to market and business opportunities (SFI Advisors, 2019).

There is a lot of scope for R&D with respect to seed development. Even these varieties of seeds are expected to serve even in unfavourable climatic conditions. For realizing maximum revenue and improving living conditions of our farming community, productivity of the crops should be improved which is possible with good management practices along with good quality of inputs. India is able to record only 50 to 60 % of the average world production per hectare. Further, there is a lot scope in the area of agro- tech products. There is a gradual shift happening from the usage of chemical intensive fertilizers and pesticides to natural manure and pesticides. This gradual shift is again opening up huge potential and opportunities for production and marketing of bio-pesticides, eco-friendly agrochemicals and natural manures. At the stage of farming, the chief objective is to maximize the output and leveraging the advantage of seasons.

Youth and Agriculture Education

Agriculture as a subject is generally opted by the students who do not get admission in medicine/engineering/management/veterinary courses to pursue their career. Therefore, talent within the available pool has to be nurtured by providing excellent teaching environment and facilities. The faculty should be adequately knowledgeable, updated, trained and highly motivated to teach the newer concepts and methodologies. The college campuses ought to be world class to offer 'first sight love' to the institution by students and ultimately to the profession. The frail infrastructure and facilities available in the educational institutions can in no way enthuse to a new entrant about the profession. On the line of IIMs, IITs, AIIMS, this is the high time to establish world class institutions of higher agricultural education in the country mainly focusing PG teaching and related research to attract and nurture the best talent and also draw more number of foreign students.

As more jobs are being created in private sector in the developing countries including India, there is a growing interest among the students from the developed countries to come and study in India to understand Indian Agriculture. Economic Survey 2017-18 says that with growing rural to urban migration by men, there is 'feminisation' of agriculture sector, with increasing number of women in multiple roles as cultivators, entrepreneurs, and labourers. The National Agricultural Research System in India is supported by agricultural higher education. The cornerstone for need-based human resources and high-quality education is a strong network of higher education institutions. It is one of the world's largest educational networks. There are public, private, and traditional universities in India's agricultural higher education system. Public agriculture universities are aided by the Indian Council of Agricultural Research (ICAR) and state governments, while private agriculture universities are supported by various government universities and University Grant Commission (UGC). There are currently 76 agricultural universities in India, including 65 state agricultural universities, 04 ICAR-deemed universities, 03 central agricultural universities and 04 central universities with agriculture faculties. These universities offer 11 disciplines for Undergraduate (UG) courses, 96 for Postgraduate (PG) courses and 73 for Doctorate (PhD) courses. In addition to these private and traditional universities also offers agriculture and allied courses like Horticulture, Forestry, Food Technology, Veterinary Sciences, Home Science etc. These SAUs are funded by the state government and ICAR special fund and central agricultural universities are supported by the ICAR. Agricultural education is regulated by the Indian Council of Agricultural Research, whereas veterinary education and forestry are regulated by the Veterinary Council of India (VCI) and the Indian Council of Forestry Research and Education (ICFRE), respectively.

National Mission on Youth in Agriculture

The Committee has emphasized for *National Mission on Youth in Agriculture* to build new skills in youth for innovative agriculture through both formal and informal education with the major objective to impart better knowledge and skill to youth on (i) sustainable, secondary and specialty agriculture, (ii) efficient knowledge dissemination, including information communication technology (ICT), (iii) technical backstopping for innovative farming, (iv) new agri-business models, and (v) entrepreneurship as well as linking farmers to markets through value chain. The committee has stressed upon the importance of imparting agricultural education right from the school level and initiation of entrepreneurship training through vocational and formal diploma programs by the CAUs, SAUs and ICAR institutes. The university curriculum also needs to be revisited to address the emerging needs and aspirations of present-day youth and markets.

Youth-Agriculture nexus

The committee has recommended the development of a new research agenda for *Youth- Agriculture Nexus* with the primary aim to (i) delineate different contexts for youth-oriented agricultural research, (ii) identify opportunities for young people's engagement in agricultural research and innovation for development (ARI4D), and (iii) determine future pathway of youth for attaining sustainable agricultural growth and income.

Plough-to-Plate initiative

The committee felt that the involvement of youth in *'Plough-to-Plate'* initiative can help in doubling farmers' income. Greater thrust is required to be given to networking for knowledge sharing/dissemination, participation of youth in out-scaling of innovations through their validation using technology parks/ innovation platforms, use of ICT, creation of agri-clinics, supporting mentoring/ hand-holding, and awareness regarding intellectual property rights (IPRs).

'Youth as a farmer' to 'Youth as value chain developer'

The committee has recommended the need for paradigm shift from 'youth as a farmer' to 'youth as value chain developer'. To provide better economic opportunities for rural youth in the changing agricultural scenario, the mindset of youth needs to move beyond the plot/ field level agriculture i.e. from production to post-production level and to link with market for better income opportunities. The combination of agricultural value chains, technology and entrepreneurship will open up tremendous economic opportunities for youth in both the farm and non-farm sectors. Therefore, they need to be encouraged to set-up agri-service centers to offer custom-hire services for small and marginal farmers for mechanizing their farm operations to enhance production at reduced cost.

Institutionalization of Incentives and Award/Reward System

To inspire and attract youth to adopt agriculture as a profession for happy living, incentives and awards/rewards need to be institutionalized. This should be a strategic priority at the local, state, and country level to ensure youth-led inclusive growth in agriculture.

Successful Entrepreneurs as Role Models for Youth

The successful entrepreneurs need to be identified and must be encouraged to act as role models for capacity development/technical back-stopping of other youth. A compendium of youth-led success stories/case studies of young agricultural entrepreneurs and innovators in various sectors of agriculture from different eco-regions of the country may be prepared on priority and made accessible to others.

Agri-Youth Innovation Corpus Fund

As part of Corporate Social Responsibility (CSR) and enhance rural employment through special projects, the private sector is required to play a proactive role in creating much needed 'Agri-Youth Innovation Corpus Fund.' This initiative would enhance rural employment opportunities through small agri-business start-ups, public-private as well as private-private entrepreneurship. Private players may also help through soft loans and mentoring programs for involving rural youths as input dealers/suppliers as well as paid extension agents.

Creation of Department of Youth in Agriculture

The committee strongly recommended the creation of a separate *Department of Youth in Agriculture* under the MoA & FW. This will ensure collaboration and coordination with concerned line departments in other Ministries such as Science and Technology, Skill Development and Entrepreneurship, Food Processing Industry, Rural Development, Commerce and Industry, Chemicals and Fertilizers, etc. so as to meet the aspirations of youth in agriculture. Such an institutional mechanism, with funding support through the proposed 'National Mission on Youth in Agriculture' will help in motivating and attracting youth in agriculture and allied fields.

ICT knowledge Enabled Youth

The committee was of the view that the role of well-trained and competent youth, with expertise in ICT application for e-NAM, Start-up, Stand-up and skill development schemes, agribusiness enterprises, etc. is extremely important. They would thus need enabling policies for long-term investments, availability of easy and soft credit, provision of subsidy upfront to the entrepreneurs, farmer exchange visits, easy market accessibility, land law reforms for entrepreneurs, no taxation system for rural-based primary value addition involving youth, review of Agri-Clinic support system by the NABARD, reforms in marketing laws such as scrapping of APMC Act, provision of ready insurance for covering risk of 'start-up' entrepreneurs, etc. would immensely encourage youth to embrace agriculture.

Conclusion

In conclusion, prioritizing skill development in agriculture is vital for fostering sustainable growth and resilience in the sector. By equipping agricultural workers with the necessary skills and knowledge, we can enhance productivity, promote innovation, and improve the quality of agricultural produce. Moreover, investing in skill development not only benefits individual farmers and workers but also contributes to broader socioeconomic development, including rural livelihoods, food security, and environmental sustainability. Governments, policymakers, and stakeholders must recognize the importance of continuous investment in education, training, and capacity-building programs tailored to the needs of the agricultural workforce. Additionally, partnerships between public and private sectors, academia, and civil society are essential for creating effective skill development initiatives and ensuring their widespread implementation. Ultimately, by empowering agricultural workers with the skills they need to thrive in a rapidly evolving sector, we can build more resilient and prosperous agricultural systems that can meet the challenges of the present and future while contributing to the well-being of communities and economies worldwide.

References

All India Survey on Higher Education (AISHE) Report, 2018-19. Department of Higher Education, Ministry of Human Resource Development, Govt. of India (http://aishe.nic. in/aishe/ view Document. action? Document Id=263).

Annual Report (2018-19). Department of Agriculture, Cooperation and Farmers Welfare, Ministry of Agriculture and Farmers Welfare, Govt. of India. (http://agricoop.nic.in/ sites/default/files/ AR_2018-19_Final_for_Print.pdf).

Annual Report (2018-19). Indian Council of Agricultural Research-Department of Agricultural Research and Education, Ministry of Agriculture and Farmers Welfare, Govt. of India. (Https:// icar.org.in/reports/DARE-ICAR-AR-19-20/DARE-ICAR-AR-2018-19-Eng% 20(2).pdf).

Bhat, P., Bhat, S. and Shayana, A. 2015. Retaining youth in agriculture: Opportunities and challenges. International Journal of Management and Social Sciences, 3.

FAO, 2007. Enabling Environments for Agribusiness and Agro-industrial development in Africa, Proceedings of a FAO Workshop Accra', Ghana.

https://www.un.org/development/desa/youth/wpcontent/uploads/sites/21/2018/12/WorldYouthReport-2030Agenda.pdf. UN, 2019. International Youth Day, 2019 Speech. New York: UN.

https://www.un.org/development/desa/youth/wpcontent/uploads/sites/21/2019/08/WYP2019_10-Key-Messages_GZ_8AUG19.pdf. UN, 2020. World Youth Report: Executive Summary. New York: UN.

Matsuyama, K. 1992. Agricultural productivity, comparative advantage, and economic growth. Journal of economic theory, 58(2): 317-334.

S.S. Khanka. Entrepreneurial Development', 2012, 104-106.

Saab, W. and Shakhovskoy, M. 2019. Pathways to Prosperity-Understanding youth's rural transitions and service needs.

World Bank. 2012. World Development Report 2013: Jobs. World Bank, Washington.

2

Impact of Women Empowerment on Growth of Indian Agriculture: Current Scenario

Devansh[1], *Anurag Tripathi*[2], *Suman Gupta*[3], *Himanshu Bhatt*[4], *Ankit Kumar Maurya*[5] and *Shudhanshu Baliyan*[6]

[1]*Department of Agricultural Economics, GBPUAT, CoA, Pantnagar, Uttarakhand*
[2]*Department of Agricultural Meteorology, GBPUAT, CoA, Pantnagar, Uttarakhand*
[3]*Department of Agricultural Economics, ANDUAT, CoA, Kumarganj, Ayodhya Uttar Pradesh*
[4]*Department of Vegetable Science, GBPUAT, CoA, Pantnagar, Uttarakhand*
[5]*Department of Agricultural Economics and Statistics, CoA, CSAUAT, Kanpur Uttar Pradesh*
[6]*Department of Entomology, GBPUAT, CoA, Pantnagar, Uttarakhand*

Abstract

Women, the integral part of society, play a significant role in sustainable development of economy through their household and other activities. Still, they are ignored; face discrimination and many obstacles in every aspect of life. Empowering women is the pivotal role in the empowerment of family and society as well. Women empowerment would lead toward improving livelihood by reducing food security because they can spend more on their families. Women must have financial liberty and role in decision making. Rural women are the major contributors in agriculture and its allied fields. Her work ranges from crop production, livestock production to cottage industry. From household and family maintenance activities, to transporting water, fuel and fodder. Despite such a huge involvement, her role and dignity has yet not been recognized. Women's status is low by all social, economic, and political indicators. Women's wage work is considered a threat to the male ego and women's engagement in multiple home-based economic activities leads to under remuneration for their work. Women spend long hours fetching water, doing laundry, preparing food, and carrying out agricultural duties. Not only are these tasks physically hard and demanding, they also rob girls of the opportunity to study. The nature and sphere of women's productivity in the labour market is largely determined by socio-cultural and economic factors. Women do not enter the labour market on equal terms when compared to men. Their occupational choices are also limited due to social and cultural constraints, gender bias in the labour market, and lack of supportive facilities such as child care, transport, and accommodation in the formal sector of the labour market. Women's labour power is considered inferior because of employers predetermined notion of women's primary role as homemakers. As a result of discrimination against female labour, women are concentrated in the secondary sector of labour market. Their work is low paid, low status, casual, and lacks potential upward mobility.

Keywords: *Women, Rural women, Women empowerment, Agriculture etc.*

Introduction

Women in India constitute nearly half of the population. They contribute 75% to the development of our society as compared to men who only 25%. According to the census 2011, nearly 98 million women in India have agriculture jobs, but around 66.1 million of them are depended on others' farms as agriculture labours. They are engaged largely in cotton, tea, oil seeds and vegetable production. It is also worth mentioning that from the basic level of work concerning farm related activities to unskilled farm jobs like sowing, transplanting, weeding, harvesting and post harvesting activities like winnowing, processing, storage etc. women are engaged right from the start.

Woman's contribution to the fisheries sector is 22% according to the FAO. Food security, horticulture, sericulture, livestock production and collection of non- timber forest are some of the allied sectors in which they are involved. They are also involved in other allied sectors which include cattle management, milk and poultry farming, and fodder collection. At the same time, women's domestic activities include cooking, child rearing water collection from the nearby and far-off places, fuel wood gathering and the overall upkeep of household. Ironically, women face disparities in terms of compensation, their share in properties and in representation in local bodies. Much of their contribution in the food production system as food producers and providers has not received effective recognition. This affects their lives and familial conditions in terms of poor health of children and low educational attainments.

"In order to awaken the people; it is the women who have to be awakened, once she is on move, the family moves, the village move, the nation moves"

-Pandit Jawaharlal Nehru.

Some historians believe that it was women who first domesticated crop plants and thereby initiated the art and science of farming. While men went out hunting in search of food, women started gathering seeds from "the native flora and began cultivating those of interest from the point of view of food, feed, fodder, fiber and fuel."

- M.S. Swaminathan

Meaning of Empowerment

Literally empowerment means the act of granting power, right, or authority to someone or something to perform various acts or duties. It is the state of being empowered to do something. Through empowerment individual acquire the power to think and act freely, exercise choice and fulfill their potential as full and equal members of society.

Role of Women in Agriculture and Its Allied Fields

Rural women perform numerous labour-intensive jobs such as weeding, hoeing, grass cutting, picking, cotton stick collections, separation of seeds from fibre. Women are also expected to collect wood from fields. This wood is being used as a major fuel source for cooking. Because of the increasing population pressure, over grazing and desertification, women face difficulties is searching of fire wood. Clean drinking water is another major problem in rural areas. Like collection of wood, fetching water from remote areas is also the duty of women. Because a rural woman is responsible for farm activities, keeping of livestock and its other associated activities like milking, milk processing, and preparation of ghee are also carried out by the women. It is common practice in the rural areas to give an animal as part of a women's dowry. Studies have revealed rural women earn extra income from the sale of milk and animals. Mostly women are engaged in cleaning of animal, sheds, watering and milking the animals.

Rural women are also responsible for collection, preparing dung cakes an activity that also brings additional income to poor families. Evidently, rural women are involved in almost all livestock related activities. Except grazing, all other livestock management activities are predominantly performed by females. Majority of women are involved in shed cleaning and collection of farm and manure. In order to generate more and more income, rural women often sell all eggs and poultry meat and leave nothing for personal use. Due to poverty and lack of required level of proteins most of women have got a very poor health. Most of women suffer from malnutrition.

Table 1: Occupational distribution of women in various sector

Type of worker	Occupational Distribution (%)	
	Women	Men
Professional	7	7
Sales	4	14
Service	7	5
Production	22	37
Agricultural	59	33
Other	2	4

In the table 1 occupational distribution by gender unveils diverse patterns in the workforce. Professional roles display parity, with women and men each comprising 7%, signaling a balanced representation. However, a notable gender gap is evident in sales, where men dominate at 14% compared to women's 4%, indicating potential disparities in career preferences or societal expectations. Service occupations witness a higher participation of women (7%) than men (5%), while production roles lean towards men (37%) over women (22%), underscoring historical norms in physically demanding work. The agricultural sector reflects a substantial gender gap, with 59% women and 33% men, highlighting cultural and sector-specific influences. These findings emphasize the need for targeted initiatives to address occupational gender disparities and foster inclusive workplaces.

The occupational distribution by gender in the agricultural sector reveals a distinctive pattern, showcasing a substantial gender gap with 59% women and 33% men engaged in agricultural work. This stark contrast may be influenced by historical norms, cultural factors, and evolving trends in the agriculture industry. The higher representation of women in agricultural roles could signify their significant contribution to this sector, potentially challenging traditional gender roles. Understanding and addressing these disparities are crucial for implementing targeted initiatives that promote gender equality in agricultural occupations, ensuring fair opportunities and recognition for all individuals contributing to this vital industry.

Schemes by Government for Women in Agriculture

Based on World Bank data, only 17.5% of India's gross product (GDP) accounted for by agricultural production. Ministry of Agriculture & Farmers Welfare provide that states and other implementing agencies to insure at least 30% expenditure on women farmers.

1. Mahila Kisan Sashaktikaran Pariyojana

Ministry of Rural Development launched a scheme namely "Mahila Kisan Sashaktikaran Pariyojana (MKSP)", as a sub component of DAY-NRLM (Deen Dayal Antodaya Yojana --- National Rural Livelihoods Mission). The "Mahila Kisan Sashaktikaran Pariyojana" MKSP a sub component of the Deendayal Antodaya Yojana-NRLM (DAY-NRLM) seeks to improve the present status of women in Agriculture, and to enhance the opportunities available to empower her. The focus of MKSP is on capacitating smallholders to adopt sustainable climate resilient agro-ecology and eventually create a pool of skilled community professionals.

Objective: Its objective is to strengthen smallholder agriculture through promotion of sustainable agriculture practices such as Community Managed Sustainable Agriculture (CMSA),Non Pesticide Management (NPM), Zero Budget Natural Farming(ZBNF), Pashu-Shakti model for doorstep animal care services, Sustainable regeneration and harvesting of Non-Timber Forest Produce.

2. Agriculture Technology Management Agency (ATMA)

Representation of Women farmers in decision making bodies-Provision for mandatory representation of women farmers in State, District, Block Farmers Advisory Committee at district level.

As Beneficiary- At least 30% of total schemes beneficiaries are to be women; and minimum 30% of resources meant farmers and women extension functionaries.

3. Agro-Clinic & Agri-Business Centers (ACABC)

Back-Ended Composite Subsidy 44% Back–ended composite subsidy towards cost of project to women as compared 36% men.

4. Mission for Integrated Development of Horticulture (MIDH)

Specific coverage of scheduled caste, Scheduled tribe, and women beneficiaries for programmatic interventions. Assistance for horticulture mechanization also available grower associations/farmer groups/ Self Help Groups having at least 10 members, who are engaged in cultivation of horticultural crops, provided the balance 60% of the cost of machines and tools is borne by such groups.

Other schemes and missions for women in agriculture

a) State Extension Programs for Extension Reforms,

b) National Food Security Mission,

c) National Mission on Sustainable Agriculture,

d) Sub – Mission for Seed and Planting Material and,

e) Sub- Mission on Agricultural Mechanization.

Why Women are Key Agent in Development of Agriculture?

With increase in urbanization, there is "feminization" of agriculture sectors with increasing number of women in multiple roles. Most rural women are engaged in agricultural practices in 3 different ways-

1. Paid laborers
2. Cultivators doing labor on their own land.
3. Managers.

They are responsible for integrated management and use of natural resources to meet the daily household needs.

Major issues faced by women in agriculture

Agriculture extension programmes ensure that information on new technologies, plant verities and cultural practices reaches farmers. However, in the developing world it is common practice to direct extension and training services primarily towards men. A recent FAO survey showed that only 5% of all agricultural extension services worldwide and that only 15% of the world's extension agents are women. To compound the problem, extension meetings are often scheduled at times when women farmers are unable to attend because of their other household responsibilities. The major issues that women face in agriculture are:

- Over-burden of work.
- Impact of technology.
- Facilities and support services.
- Development bias.
- Constraints to women's access to resources.
- Access to land.

- Access to credit.
- Access to markets.
- Research and technology development.
- Access to extension and training.

Indian Women Farmers & Farming who will Inspire you to Rally for Agriculture

Spinning a Destiny from Cotton: Atram Padma Bai

She setup a Hiring Centre for agriculture tools to lend farming tools like pickaxes, sickles, spades, hoes and wheelbarrows to poor farmers at marginal rates. And after 6 years, she has certainly come a long way.

Today, she is Sarpanch to eight villages and two thousand farmers. She has built several fair weather and concrete roads and acquired government sanctions to make clean water accessible and build reservoirs to harvest rainwater. In a world where most women farmers have no claim over their land. Padma has truly built a life on her terms.

Leading Women Farmer, Mushrooming Entrepreneurship Spirit; Bina Devi

Recipient of Nari Shakti Puraskar in 2020. Bina Devi's work in promoting fungi culture or mushroom cultivation awarded her the title "Mushroom Mahila". This teacher – turned – businesswomen from Tilkari village in Bihar is credited with promoting self- employment among rural women.

Besides helping 105 neighboring villages take up mushroom cultivation. Bina Devi has actively promoted digital literacy, organic vermicompost production & insecticide preparation amongst rural women farmers. Her one – women – movement helped 2,500 farmers learn and adopt climate – smart practices like the system of rice intensification.

The Fortune- telling Women Farmer of Muzaffarpur: Rajkumari Devi

Rajkumari Devi can't predict the future. What she can predict is what your soil yields best. Hailing from Anandpur village in Muzaffarpur. She has rewritten the agriculture destiny of 19 other villages in her district. It isn't uncommon to see her inspecting fields on request to access everything from soil quality to predicting its harvests.

Rajkumari Devi has been the driving force behind women in the region not only farming but becoming full- fledged entrepreneurs. Today, she leads

Anandpur jyoti center – a non- profit that employs women farmers to make products like jam, jellies and pickles from their produce.

Women supervising the agricultural operations performed by the laborers:

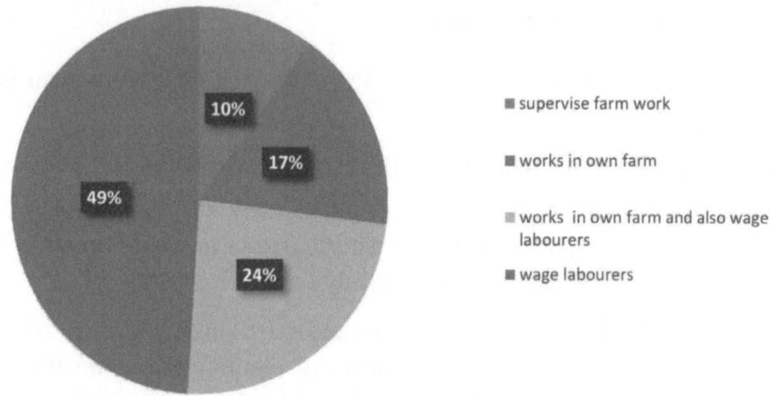

Dimensions of Women Empowerment in Agriculture

- Input in product decisions and autonomy in production
- Control over use of income
- Access to and decisions on credit
- Work load and time allocation
- Education

Recommendations

For the recognition of women contribution in agriculture and its allied fields and reducing the gender issues, these are the following recommendations:

- Recognition of labour work of working women in the rural economy may be accounted in monetary terms.
- More facilities should be provided to poor rural women for land, agricultural and livestock extension services.
- Priority must be given to women in accessing credit on soft terms from banks and other financial institutions for setting up their business, for buying properties, and for house building.
- Review, revision and reformation of cooperative legislation and government policies which facilitate and encourage women to become

members of cooperatives and participate in decision-making processes.

- Measures should be taken to enhance women's literacy rates. A separate education policy for women may serve the purpose.
- Identification of an appropriate mechanism which could provide development opportunities to women in rural areas.
- Encouraging cooperatives to have special programmes and tasks for women to perform in the organisational and business affairs. It has been observed that in many of the countries region more women are being taken in to undertake administrative and functional activities-they make very good, reliable and honest cashiers, sales girls, inventory controllers, secretaries, public relations officers and member contract persons.

Conclusion

Rural women are the major contributors in agriculture and its allied fields. Her work ranges from crop production, livestock production to cottage industry. From household and family maintenance activities, to transporting water, fuel and fodder. Despite such a huge involvement, her role and dignity has yet not been recognized. Women's status is low by all social, economic, and political indicators. Women's wage work is considered a threat to the male ego and women's engagement in multiple home-based economic activities leads to under remuneration for their work. Women spend long hours fetching water, doing laundry, preparing food, and carrying out agricultural duties. Not only are these tasks physically hard and demanding, they also rob girls of the opportunity to study. The nature and sphere of women's productivity in the labour market is largely determined by socio-cultural and economic factors. Women do not enter the labour market on equal terms when compared to men. Their occupational choices are also limited due to social and cultural constraints, gender bias in the labour market, and lack of supportive facilities such as child care, transport, and accommodation in the formal sector of the labour market. Women's labour power is considered inferior because of employers predetermined notion of women's primary role as homemakers. As a result of discrimination against female labour, women are concentrated in the secondary sector of labour market. Their work is low paid, low status, casual, and lacks potential upward mobility.

References

Anon., Census of India. 2001, Office of Registrar General and Census Commissioner, Govt of India, New Delhi, 2004.

Anon., Census of India. 2011, Office of Registrar General and Census Commissioner, Govt of India, New Delhi, 2013.

Dietz, T., Estrella Chong, A., Font Gilabert, P., & Grabs, J., 2018. Women's empowerment in rural Honduras and its determinants: Insights from coffee communities in Ocotepeque and Copan. Development in Practice, 28(1): 33–50.

Diiro, G. M., Seymour, G., Kassie, M., Muricho, G., & Muriithi, B. W., 2018. Women's empowerment in agriculture and agricultural productivity: Evidence from rural maize farmer households in western Kenya. PLoS One, 13(5): 197-200.

FAO, 2011. The State of Food and Agriculture- Women in Agriculture Closing the gender gap for development. Rome.

Gite, L. P., 2009, Women in Indian agriculture. In Ergonomics in Developing Regions: Needs and Applications (ed. Scott, P. A.), CRC Press: 291–306.

Gite, L. P., Majumder, J., Mehta C. R. and Khadatkar, A., 2009. Anthropometric and strength data of Indian agricultural workers for farm equipment design. Book No. CIAE/2009/4, ICAR-Central Institute of Agricultural Engineering, Bhopal.

ICAR-CIAE, 2014, Vision 2050, CIAE Bhopal

ICAR-CIWA. Drudgery reduction and women friendly farm tools and equipment's, Bhubaneswar, 2010.

ICAR-CIWA. Strengthening gender perspective in agricultural research & extension. Bhubaneswar, 2015.

ICAR-CIWA. Vision 2050, Bhubaneswar, 2015.

Maiti, R. and Ray, G.G., 2004. Manual lifting load limit equation for adult Indian women workers based on physiological criteria. Ergonomics, 47(1): 59–74.

Mehta, C. R., Chandel, N. S. and Senthilkumar, T., 2014. Status, challenges and strategies for farm mechanization in India. Agricultural Mechanisation in Asia, Africa and Latin America (AMA), 45(4): 43–50.

Narayan-Parker, D., 2005. Measuring Empowerment: Cross-disciplinary Perspectives. Washington, DC: World Bank.

Singh, S. P., Gite, L. P., Majumder, J. and Agarwal, N., 2008. Aerobic capacity of Indian farm women using sub-maximal exercise technique on tread mill. Agric. Eng. Int.: CIGR.

3

Millets Nutritional Powerhouses with Health Benefits

***Harsh Gupta*[1]*, Umesh Patle*[2]*, Ankit Soni*[3] *and Rahul Kumar Rai*[4]**

[1,2&3]*MGCGVV, Facultyof Agriculrue Science Chitrakoot, Satna, Madhya Pradesh*
[4]*Department of Agtricultural Economics, CoA, BUAT, Banda, Uttar Pradesh*

Abstract

Millet is a type of grain that plays a significant role in the diets of many regions, particularly in areas with limited water. It's a good choice for agriculture because it can withstand drought conditions and produce well. Millets also offer valuable nutrition. They contain phytochemicals, which are natural compounds that can be beneficial for health. However, the types and amounts of these phytochemicals can vary between different kinds of millet. The way millet is processed can affect the amount of these beneficial phytochemicals. Methods like dehulling, decortication, malting, fermentation and heating can reduce the levels of these compounds. As a result, millet-based-foods and drinks tend to have lower levels of phytochemicals compared to other cereal grains. Apart from this, there is some evidence to suggest that millet-based-foods and beverages can offer health benefits, including anti-diabetic, anti-obesity and cardiovascular benefits. These effects are attributed to the phytochemicals in millet, which can also support the immune system. However, it's important to note that most research has been conducted in labs or on extracts from millet, not in real people. The goal of this chapter is to gather existing information on the nutritional value and health benefits of millet. This information could encourage nutritionists and dietitians to promote the consumption of millets in people's diets, considering their nutritional profile and the contribution of phytochemicals.

Keywords: *Health benefits, Millet-based-foods, Millet, Nutritional and Phytochemicals, etc.*

Introduction

Millet is a very important cereal grain that many people around the world eat. In fact, over a third of the world's population consumes millet. It's the sixth most produced cereal crop globally. Some common types of millet include Jowar (Sorghum), Sama (Little millet), Ragi (Finger millet), Korra (Foxtail millet), and Variga (Proso millet). Among these, Bajra and Sama have more fat, while Ragi has the least. Millet has been cultivated for thousands of years and is used in many parts of the world. In the past, even the Romans and Gauls enjoyed millet-based porridges. Today, China, India, Greece, Egypt, and various African countries are the major producers of millet. In rural areas, some types of millet, like Finger millet and Sorghum, are eaten by people, while others are primarily used as animal feed. These millets have packed with several beneficial nutrients and play a crucial role in traditional diets in many regions. Compared to commonly used rice and wheat, millet is three to five times more nutritious. While wheat and rice mainly provide food security, millet offers not only food security but also health benefits and livelihood opportunities. It can help manage health issues like diabetes and high cholesterol.

In India, Karnataka is the top producer of millets, accounting for over 58% of global production. Despite its nutritional and health benefits, many people in India are not aware of the advantages of consuming millet. (Upadhyaya *et al.,* 2007). Millets are fantastic crops when it comes to pest management. Traditionally grown millets don't require pesticides and the land used for millet cultivation remains largely pest-free. Some millets, like foxtail millet, not only resist pests themselves but also serve as natural protectors against pests in storage conditions for other crops like green gram. This means there's no need for fumigants to keep them safe. It has a treasure trove of micronutrients, including vitamins and beta-carotene, which are now seen as essential as pharmaceutical pills. In the present day, millets are exceptionally superior and serve as a solution to the widespread issues of malnutrition and obesity affecting a significant portion of the Indian population (Singh KP *et al.,* 2012). Millets are rich in essential fatty acids, including linoleic, oleic, and palmitic acids, which are present in their free form. They also contain compounds like monogalactosyl, diacylglycerols, digalactosyl diacylglycerols, phosphatidylethanolamine, phosphatidylserine and phosphatidylcholine in bound forms. Although other fatty acids like arachidic acid, behenic acid and erucic acid are found in millets, they are in trace amounts. Millet oil can be a valuable source of linoleic acid and tocopherols. Additionally, millets are gluten-free and have an alkaline effect in the body. They contain essential B vitamins such as niacin, folacin, riboflavin, and thiamine, as well as phosphorus. These nutrients play a crucial role in energy synthesis within the body.

Nutritional Profile of Different Millets and other Cereals

Table 1: Nutrient composition of millets and other cereals (per 100 g edible portion; 12% moisture)

Food	Protein*(g)	Fat (g)	Ash (g)	Crude fiber (g)	Carbohydrate (g)	Energy (kcal)
Rice (brown)	7.9	2.7	1.3	1.0	76.0	362
Wheat	11.6	2.0	1.6	2.0	71.0	348
Maize	9.2	4.6	1.2	2.8	73.0	358
Sorghum	10.4	3.1	1.6	2.0	70.7	329
Pearl millet	11.8	4.8	2.2	2.3	67.0	363
Finger millet	7.7	1.5	2.6	3.6	72.6	336
Foxtail millet	11.2	4.0	3.3	6.7	63.2	351
Common millet	12.5	3.5	3.1	5.2	63.8	364
Little millet	9.7	5.2	5.4	7.6	60.9	329
Barnyard millet	11.0	3.9	4.5	13.6	55.0	300
Kodo millet	9.8	3.6	3.3	5.2	66.6	353

*All values except protein are expressed on a dry weight basis

Sources: Hulse and others (1980); FAO (1995)

Table 2: Relative properties of vitamin, macro and micro nutrient content of different millets and other cereals

Food	Ca (mg)	Fe (mg)	Thiamin (mg)	Riboflavin (mg)	Niacin (mg)
Rice (brown)	33	1.8	0.41	0.04	4.3
Wheat	30	3.5	0.41	0.10	5.1
Maize	26	2.7	0.38	0.20	3.6
Sorghum	25	5.4	0.38	0.15	4.3
Pearl millet	42	11.0	0.38	0.21	2.8
Finger millet	350	3.9	0.42	0.19	1.1
Foxtail millet	31	2.8	0.59	0.11	3.2
Common millet	8	2.9	0.41	0.28	4.5
Little millet	17	9.3	0.30	0.09	3.2
Barnyard millet	22	18.6	0.33	0.10	4.2
Kodo millet	35	1.7	0.15	0.09	2.0

Sources: Pulse and others (1980); FAO (1995)

Health Benefits of Millets

1. Millets and Diabetes

Millets have shown promise in managing diabetes by reducing α-glucosidase and pancreatic amylase activities, thereby lowering postprandial hyperglycemia. They inhibit the enzymatic breakdown of complex carbohydrates. Millets also impact enzymes like aldose reductase, which is involved in preventing the accumulation of sorbitol and reducing the risk of diabetes-related cataracts. Consuming millets helps control blood glucose levels and aids in dermal wound healing through antioxidants (Rajasekaran NS, *et al.*, 2004).

In collaboration with the Indian Institute of Millets Research, Hyderabad, the National Institute of Nutrition (ICMR) assessed the Glycemic Index (GI) of sorghum-based foods in 2010 under the National Agricultural Innovation Project (NAIP). The results revealed that sorghum-based foods have a low GI, which lowers postprandial blood glucose levels. Finger millet diets exhibit a low glycemic response due to their high fiber content and are linked to dermal wound healing. Studies provide strong evidence for the role of millet proteins in inhibiting cataractogenesis in humans. Currently, diabetes is a widespread health issue worldwide, and millets play a preventive role in Type II Diabetes due to their significant magnesium levels. Magnesium is a crucial mineral that enhances insulin efficiency and glucose receptor function by promoting the production of carbohydrate-digesting enzymes, which ultimately helps manage insulin action (O.S.K. Reddy, 2017).

2. Millets and Obesity

Obesity is a growing concern in India, and it's linked to various chronic diseases such as diabetes and cardiovascular issues. Recent research indicates that a diet rich in dietary fiber can lower the risk of obesity (Alfieri *et al.*, 1995). Foods with high dietary fiber content enhance bowel function, slow down the digestion and absorption processes, and consequently reduce the risk of chronic diseases (Ali *et al.*, 1982). Millets contain a substantial 22% dietary fiber, which is significantly higher than other cereals like wheat (12.6%), rice (4.6%), and maize (13.4%). For instance, finger millet consists of 15.7% insoluble dietary fiber and 1.4% soluble dietary fiber (Chethan *et al.*, 2007). Additionally, finger millet boasts a total dietary fiber content of 22.0%, with 19.7% insoluble and 2.5% soluble dietary fiber (Shobana *et al.*, 2007).

The dietary fiber in millets can be classified into soluble and insoluble fibers. Obesity is increasingly prevalent and is associated with diabetes, high blood pressure, and heart problems. Studies suggest that a high-fiber diet improves bowel function and lowers the risk of obesity by enhancing digestion and absorption in the body, consequently reducing the risk of chronic diseases. Millets help in satiating hunger, promoting weight management, and reducing obesity. With their high fiber content, millets also alleviate issues like constipation, flatulence, bloating, and stomach cramping. Good digestion and absorption further reduce the likelihood of gastrointestinal illnesses such as ulcers and colon cancer (O.S.K. Reddy, 2017).

3. Millets and Cancer

Studies have revealed that millets are rich in phenolic acids, phytates, and tannins, which are considered antinutrients with the potential to reduce the risk of colon and breast cancer. Phenolics in millets have demonstrated effectiveness in preventing cancer initiation and progression in vitro (Chandrasekara A, *et al.*, 2011).

Sorghum, another type of millet, has well-documented anti-carcinogenic properties. The polyphenols and tannins present in sorghum possess anti-mutagenic and anti-carcinogenic properties. They can act against human melanoma cells and exhibit positive melanogenic activity. Studies conducted in China and various parts of the world have shown that the incidence of esophageal cancer was lower in communities where sorghum consumption was prevalent. These findings were based on a study of 21 communities over six years, revealing lower mortality from esophageal cancer in sorghum-consuming regions compared to areas where wheat and corn were more commonly consumed.

Many antioxidants found in millets not only neutralize free radicals that can lead to cancer but also help detoxify the body, particularly the kidneys and liver. Compounds like quercetin, curcumin, ellagic acid, and various other catechins promote the elimination of foreign agents and toxins while enhancing enzymatic activity in these organs (O.S.K. Reddy, 2017).

4. Millets and Phytochemicals

Millets are a rich source of phytochemicals and micronutrients, including phenolics, sterols, lignans, inulin, resistant starch, β-glucan, phytates, tocopherol, dietary fiber, and carotenoids. Among these, polyphenols, phenolic acids, tannins, and flavonoids act as antioxidants and support the body's immune system (Chandrasekara A, *et al.*, 2010).

These antioxidants not only neutralize free radicals that can cause cancer but also aid in detoxifying the body, especially the kidneys and liver. Compounds like quercetin, curcumin, ellagic acid, and various catechins help eliminate foreign agents and toxins, enhancing proper excretion and enzymatic activity in these organs. Consequently, polyphenols have garnered significant attention due to their roles in promoting human health. Millets are exceptional foods that are known for their ease of digestion and suitability for individuals with gluten sensitivities. They are rich in essential amino acids, fatty acids, and dietary fiber beneficial for the health.

1. Millets are packed with essential nutrients that are vital for the body's proper functioning. They contain minerals like iron and copper, which are necessary for blood cell production and improved blood oxygenation. Phosphorus in millets helps regulate blood pressure, enhancing the body's defense against diseases.
2. Consumption of millets in significant quantities helps reduce triglycerides in the body, preventing blood platelet clumping and lowering the risk of coronary artery diseases.
3. Millets are a source of essential B vitamins that aid in the breakdown of carbohydrates and fats. They play a role in lowering homocysteine levels in the blood, preventing the bonding of cholesterol and plaque formation. Niacin increases the "good" HDL cholesterol levels and protects blood vessels from hemorrhage and atherosclerosis.
4. Millets have a protein structure similar to wheat but without the gluten. This makes millets a preferred choice for vegans and vegetarians, as they offer excellent plant-based protein without saturated fats.
5. Millets contain tryptophan, an amino acid that helps maintain a healthy body weight by reducing appetite. Their high fiber content promotes quick satiety and prevents overeating, making them beneficial for weight management.
6. The combination of fiber and phytonutrients in millets reduces the risk of colon cancer. Lignans found in millets can also help protect against breast cancer by converting into mammalian lignan.
7. The presence of magnesium in millets relaxes the muscles of arterial walls, contributing to lowered blood pressure. Magnesium can also reduce the severity of asthma and the frequency of migraines.
8. Millets are an ideal choice for individuals with celiac disease, as they are gluten-free and well-tolerated by those who cannot digest gluten.

Conclusion

In conclusion, millets are a remarkable and highly nutritious group of cereal grains with a rich history and significant potential for promoting human health, which play a vital role in traditional diets in various regions and are known for their resilience in harsh agricultural conditions, making them a valuable asset for food security. Millets are not only rich in essential nutrients such as minerals, vitamins, and dietary fiber but also offer a range of health benefits. Their role in managing conditions like diabetes and obesity is noteworthy, with evidence pointing to their positive impact on blood glucose levels and weight management. Millets also exhibit anti-cancer properties, thanks to their abundance of antioxidants, which can neutralize free radicals and detoxify the body. The presence of phytochemicals in millets further enhances their health-promoting potential. While research has predominantly been conducted in vitro, the data collected underscores the importance of these grains in our diets. To fully harness the nutritional and health benefits of millets, it is crucial to raise awareness among nutritionists and dietitians, encouraging the public to incorporate these grains into their daily meals. Millets are not only a source of sustenance but also a pathway to improved health and well-being.

References

Chandrasekara, A. and Shahidi, F. 2010. Content of insoluble bound phenolics in millets and their contribution to antioxidant capacity. Journal of agricultural and food chemistry, 58(11): 6706-6714.

Chandrasekara, A. and Shahidi, F. 2011. Antiproliferative potential and DNA scission inhibitory activity of phenolics from whole millet grains. Journal of Functional Foods, 3(3): 159-170.

Hulse, J.H., Laing, E.M. and Pearson, O.E. 1980. Sorghum and the millets: their composition and nutritive value. Academic press.

Rajasekaran, N.S., Nithya, M., Rose, C. and Chandra, T.S. 2004. The effect of finger millet feeding on the early responses during the process of wound healing in diabetic rats. Biochimica et Biophysica Acta (BBA)-Molecular Basis of Disease, 1689(3): 190-201.

Reddy, O.S.K., 2017. Smart millet and human health. Green Universe Environmental Services Society.

Singh, K.P., Mishra, A. and Mishra, H.N. 2012. Fuzzy analysis of sensory attributes of bread prepared from millet-based composite flours. LWT-Food Science and Technology, 48(2): 276-282.

Taylor, J.R. 2019. Sorghum and millets: Taxonomy, history, distribution, and production. In Sorghum and millets: 1-21. AACC International Press.

Upadhyaya, H.D., Gowda, C.L.L. and Reddy, V.G. 2007. Morphological diversity in finger millet germplasm introduced from Southern and Eastern Africa. Journal of SAT Agricultural Research, 3(1): 1-3.

4

Bio-inoculants A Key Component for Sustainable Horticulture

***Purnita Raturi*[1], *Deeksha Semwal*[2] and *Rahul Kumar Rai*[3]**

[1]*Department of Horticulture, CoA, GBPUAT, Pantnagar, Uttarkhand*
[2]*Department of Food Science & Technology, CoA, GBPUAT, Pantnagar Uttarkhand*
[3]*Department of Agricultural Economics, CoA, BUAT, Banda, Uttar Pradesh*

Abstract

The rise in food demand sparked the "green revolution" with chemical solutions, but their harm to nature and health raised concerns. Seeking eco-friendly options led to the rise of bioinoculants in agriculture. Vital for doubling food production by 2050, these agents directly aid plants by improving nutrient uptake and root growth, while indirectly combating pathogens and stress. Sustainable farming prioritizes high yields with minimal environmental impact. Leveraging beneficial microorganisms like Plant Growth-Promoting Rhizobacteria (PGPR) and Arbuscular Mycorrhizal Fungi (AMF) is crucial. Blending traditional wisdom with modern science promises resilient, eco-conscious agriculture for generations to come.

Keywords: *Bio-inoculants, PGPR, AMF, Sustainable, etc.*

Introduction

In the middle of the 1900s, the world experienced a big increase in people, leading to a greater need for food. This demand for more food pushed for a farming revolution called the "green revolution." This revolution used chemicals to help grow more crops and fight plant diseases. But using lots of these chemicals caused problems for the environment and people's health. Even with all these chemicals, a large amount of food (up to 25% of what's grown) is still lost to diseases each year. By 2050, we might need to double how much food we make. To do this without harming the environment or health, we're looking for new ways. Some ideas include making plants that are genetically modified, creating plant varieties that resist pests naturally, and using helpful microorganisms like certain bacteria and fungi. These helpful organisms can be used in different ways, like making soil healthier or protecting plants.

The microorganisms colonizing plant roots generally include bacteria, algae, fungi, protozoa and actinomycetes. Enhancement of plant growth and development by application of these microbial populations is well evident. Among different microbial populations present in the rhizosphere, bacteria are the most abundant microorganisms. Various genera of bacteria, *Pseudomonas, Enterobacter, Bacillus, Variovorax, Klebsiella, Burkholderia, Azospirillum, Serratia and Azotobacter*, cause a pronounced effect on plant growth and are termed as plant growth promoting rhizobacteria (PGPR). PGPR play a significant role in enhancing plant growth and development both under non-stress and stress conditions by a number of direct and indirect mechanisms. Microbes exert key functions in ecosystems being involved in nitrogen fixation, phosphorus solubilization, soil formation, and production of siderophores, plant growth regulators and organic acids as well as protection by enzymes like ACC-deaminase, chitinase and glucanase.

In addition to bacterial population, fungi also represent a significant portion of soil rhizosphere microflora and influence plant growth. The symbiotic association generated by fungi with plant roots (mycorrhizae) increases the root surface area, and therefore enables the plant to absorb water and nutrients more efficiently from large soil volume. The mycorrhizal association not only increases the nutrient and water availability, but also protects the plant from a variety of biotic and abiotic stresses.

Plant Growth Promoting Rhizobacteria (PGPR): An Overview

Plant growth promoting rhizobacteria are important to horticulture production due to their ability to improve plant growth and increase plant protection from diseases and abiotic stress like drought and salinity. PGPR are the microorganisms associated with the plant's rhizosphere. They are the important contributors towards enhanced soil fertility and productivity. As a biostimulant, PGPR stimulate natural processes that increase plant nutrient uptake and efficiency, abiotic stress tolerance and quality of crops (Calvo *et al.*, 2014). Plant growth promoting bacteria have also been identified as a potential bioorganic component of nano-biofertilizers which can increase plant growth and inhibit the growth of parasitic fungi (Gouda *et al.*, 2018). They minimize the adverse effects originated from biotic as well abiotic factors. Moreover, they help in the removal of the agrochemicals and the xenobiotics from the agroecosystems. Several studies have been done to assess the beneficial effects of PGPR on the different crops including horticultural crops. The potential PGPR are *Azospirillum, Azotobacter, Bacillus, Pseudomonas,* and *Rhizobium.* Application of PGPR has shown a significant reduction in the need of chemical fertilizers and other chemical amendments. Following major mechanisms are employed by the microorganisms to promote plant growth and development. Bacteria interact with plants both above and below the ground. They can live inside the plant and around its roots, offering beneficial effects known as plant growth promotion.

Beneficial Actions of Plant Growth-Promoting Rhizobacteria (PGPR)

Plant growth-promoting rhizobacteria (PGPR) have been extensively studied for their positive effects on plant health and growth. These microorganisms, play a significant role in enhancing plant growth through direct and indirect mechanisms, which are important for their application as bio-inoculants in horticulture.

Direct Mechanisms

Through these mechanisms, PGPR provide various advantages to horticultural crops, facilitating the absorption of both micronutrients and macronutrients.

1. **Act as Potassium-solubilizing and phosphate-solubilizing bacteria**: Potassium-solubilizing bacteria (KSB) and phosphate solubilizers are types of Plant Growth-Promoting Bacteria (PGPB) crucial for enhancing soil fertility and plant growth. KSB release organic acids such as oxalic, gluconic, tartaric, citric, and others to dissolve potassium from minerals, making it more available for plants. Meanwhile, phosphate solubilizers like Bacillus and Pseudomonas produce enzymes such as acid phosphatases to convert insoluble phosphorus into forms that plants can absorb, promoting nutrient uptake and boosting plant growth. These bacteria not only improve phosphorus availability but also aid in the solubilization of other essential elements like zinc, contributing to enhanced soil fertility and improved crop productivity.

2. **Nitrogen Fixation:** Nitrogen is another crucial nutrient for plant growth, especially for crops producing fruits and seeds. Some plants, particularly legumes, have a fascinating relationship with soil bacteria called rhizobia. These bacteria can fix nitrogen from the atmosphere and convert it into a usable form (ammonia) for plants. This happens in structures called nodules, where rhizobia reside within the plant tissues. Beyond nitrogen-fixing abilities, certain non-nitrogen-fixing bacteria,

like some Pseudomonas species, can enhance the symbiotic relationship between legumes and rhizobia. This collaboration improves nitrogen fixation, benefiting plant growth.

3. **Siderophore Production:** Plant Growth-Promoting Bacteria (PGPR) play a crucial role in aiding plant iron uptake by producing siderophores—powerful compounds that bind to iron in the soil, making it available for plants. Iron is essential for various plant processes like photosynthesis and metabolism, yet its availability is limited in neutral pH soils. Bacteria like Pseudomonas are notable siderophore producers, generating compounds such as pyochelin, pseudobactin, and pyoverdine. These siderophores form complexes with iron, facilitating its absorption by plants like Arabidopsis thaliana, ultimately promoting improved growth.
4. **Production of Growth Regulators:** Certain bacteria influence plant growth by producing compounds such as auxins (e.g., IAA), gibberellins, cytokinins, and organic substances like acetoin, 2,3-butanediol, and N,N-di methyl hexa decylamine. These compounds stimulate various growth processes in plants, including seed germination, root formation, and photosynthesis. Specifically, IAA plays a pivotal role in overall plant development, while other substances contribute to positive plant-bacteria interactions, fostering plant growth.

Indirect Mechanisms of Plant Growth-Promoting Bacteria (PGPB)

PGPR cannot directly plants growth but they also indirectly support plant health by fighting off potential threats from harmful pathogens. There are some different way PGPR act with plants

1. **Antagonism towards Pathogens:** PGPR possess a variety of compounds and enzymes that can hinder or eliminate pathogens. For instance, *Pseudomonas* bacteria produce siderophores that can bind to available iron in the environment, making it less accessible to pathogens. This method restricts the growth of pathogens by limiting their access to essential nutrients, which was one of the first mechanisms discovered in PGPR.
2. **Enzymatic Attack on Pathogens:** Bacterial endophytes with antifungal properties use enzymes like chitinases, cellulases, and β-1,3-glucanases to break down the cell walls of fungi. Chitinase, for example, breaks down chitin, a significant component of fungal cell walls. Bacteria like *Bacillus species*, including *B. licheniformis, B. cereus, B. subtilis*, and *B.*

thuringiensis, produce chitinase, showing their role in controlling plant pathogens like *Botrytis cinerea*, responsible for gray mold in various plants. Similarly, other bacteria like *Pseudomonas* use enzymes to attack fungal cell walls.

3. **Production of ACC-deaminase:** PGPR strains not only tolerate stress but also promote plant growth in stressful environments. They achieve this through various mechanisms like reducing stress-induced ethylene levels, producing exopolysaccharides, and inducing systemic resistance. Certain PGPRs possess ACC deaminase activity, which lowers ethylene levels, alleviating its inhibition of plant growth and enhancing stress resistance through phytohormone signaling pathways. This positively impacts plant growth, particularly in combating abiotic stresses like drought.
4. **Induction of Plant Defense Responses:** Bacterial endophytes stimulate the plant's defense mechanisms. These responses, triggered by PGPR, help plants defend against pathogen attacks and abiotic stress. By activating various defense actions and increasing production of osmolytes, antioxidant and secondary metabolites by these bacteria, assist plants in protecting themselves.

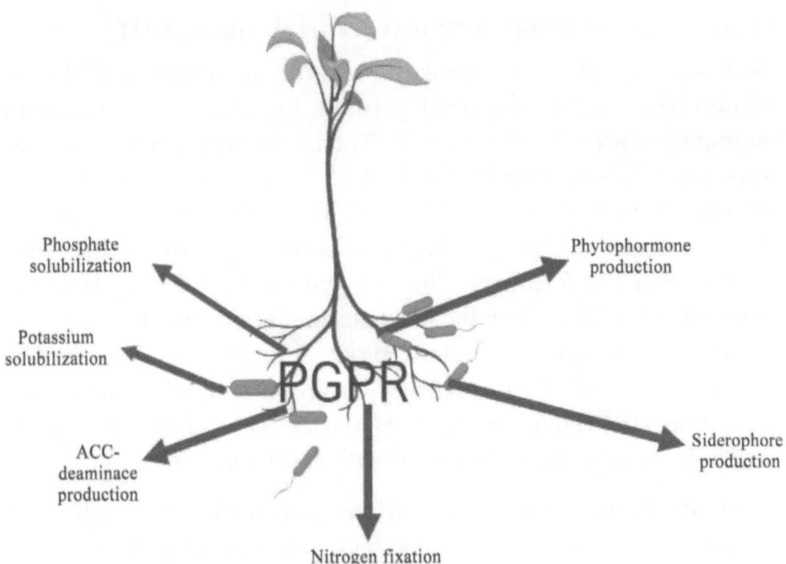

Fig. 1: Plant growth-promoting roles of PGPR

Arbuscular Mycorrhizal Fungi (AMF): An Overview

Mycorrhiza is a symbiotic association between plant roots and soil fungi, improving soil properties and acting as a barrier against foreign organisms. Fungi gain nutrients from host plants, while plants receive nutrients and water from fungi. Arbuscular Mycorrhizal Fungi (AMF) are the most common and establish connections with 80–90% of land plants, aiding various growth stages. There are two main types: Ectomycorrhiza, forming a sheath around roots, and Endomycorrhiza, which penetrates roots and creates nutrient exchange structures within root cells (arbuscules, vesicles). Vesicular Arbuscular Mycorrhiza (VAM) is a significant type, widely utilized in horticulture for enhancing crop growth.

Mode of action

Arbuscular Mycorrhizal Fungi (AMF) create a partnership between plants and fungi. Plants provide sugars via photosynthesis, while fungi deliver water and soil-derived nutrients, like phosphorus, to the plants. VAM fungus hyphae extend beyond the depletion zone, reaching distant nutrient supplies in narrow soil pores. Their extensive growth around roots maximizes nutrient uptake, expanding the nutrient-depleted area, eventually replenished by nutrient diffusion.

Beneficial actions of arbuscular mycorrhizal fungi (AMF)

1. **Nutrient uptake:** Arbuscular Mycorrhizal Fungi (AMF) act as biofertilizers, enhancing plant nutrition by retrieving and transporting nutrients, especially phosphorus (P) and zinc (Zn), from areas beyond root reach. This partnership significantly boosts nutrient uptake and water absorption in host plants. AM fungi efficiently scavenging phosphorus from vast soil volumes, swiftly transporting it to root cells, bypassing direct uptake mechanisms. In this symbiosis, the fungus exchanges nutrients for carbon from the plant, sustaining its growth and supporting spore development for future plant colonization. The extraradical mycelium (ERM) absorbs nutrients from the soil, while the intraradical mycelium (IRM) releases nutrients into the plant's interfacial apoplast in exchange for carbon to fuel its growth and functions.

2. **Role in Biotic Stress Alleviation:** Arbuscular Mycorrhizal Fungi (AMF) play a crucial role in combating plant pathogens by employing multiple defense mechanisms. They create a physical barrier, reinforce cell walls, and induce the production of compounds in host roots that resist pathogen invasion. AMF produce antifungal and antibacterial substances, compete for nutrients, stimulate beneficial microbial activity,

and foster competitive interactions in the root zone, hindering pathogen access. Additionally, they enhance antagonistic microbes, compensate for root damage, and alter root exudates, influencing pathogen interactions. AMF's diverse strategies bolster plant defense mechanisms, significantly protecting plants against various biotic stresses.

3. **Role in Abiotic Stress Alleviation:** The symbiotic relationship between plants and Arbuscular Mycorrhizal Fungi (AMF) is crucial in alleviating various abiotic stresses such as metal toxicity, salinity, and drought. AM fungi enhance antioxidant defense mechanisms in plants, combatting oxidative stress caused by reactive oxygen species (ROS). During drought, AM fungi's hyphal network extends beyond root zones, increasing water uptake and expanding the depleted water area. In salt stress, mycorrhizal roots improve water and stomatal conductance, aiding in osmotic adjustment and antioxidant production. With heavy metal toxicity, AM fungi immobilize and accumulate metals in their mycelia, mitigating toxicity effects. Additionally, glomalin proteins produced by these fungi contribute to soil stabilization and reduce metal bioavailability. Overall, AM fungi support plants in adapting to and thriving amidst various environmental challenges by enhancing stress tolerance mechanisms.

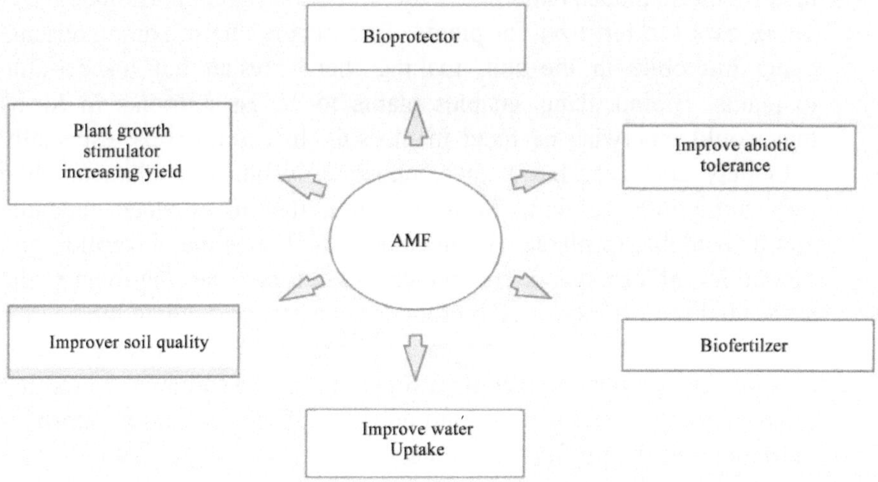

Fig. 2: Plant growth-promoting roles of AMF

Beneficial effects of bio-inoculants (AMF and PGPR) in horticulture crops

1. **Enhances plant establishment and survival:** Optimal rooting is crucial for the establishment and survival of horticulture crops, especially as they are mostly propagated vegetatively. Soil erosion in orchard ecosystems poses a challenge, depleting microbial propagules necessary for plant survival. To address this, inoculating plantations with bio-inoculants in the rhizosphere is necessary. This practice, common in litchi orchards, involves using soil from older orchards to establish mycorrhizal associations with litchi roots, promoting rapid plant growth post-inoculation. Bioinoculants also contribute to hormone production, stimulating rooting and shooting processes and enhancing plantlet survival. Overcoming the challenge of hardening micropropagated plantlets involves including bioinoculants in the hardening media, leading to improved acclimatization, survival, and growth in various micro propagated horticulture species like strawberry and banana. Bioinoculants play a crucial role in optimizing plant performance, buffering stress during acclimatization, and enhancing overall plant and soil health.

2. **Increase uptake of nutrients:** Nutrient deficiency is a common issue in horticulture plants, particularly for N, P, K, Zn, and Fe. Bioinoculants serves as a bio fertilizer for plants. This occurs due to some nutrients being immobile in the soil, making them present but inaccessible to plants. Bioinoculants enables plants to access nutrients in forms that would otherwise be fixed in the soil. In certain soils, especially extremely acidic or basic ones, phosphorus binds to elements like iron, aluminum, calcium, or magnesium, making it water-insoluble and unavailable to plants. Bioinoculants facilitates the absorption and dissolution of various organic matter and nutrients, storing them in the associated root. It increases the root surface area, extending hyphae for improved water and nutrient uptake. The use of biofertilizer enhances root branching, and connect plant roots to a wider soil area, increasing the absorption zone for water and nutrients. This association improves nutrient absorption efficiency in plants.

3. **Increase resistance to biotic and abiotic stresses:** Bioinoculants serve as bioprotectors for plants, countering various pathogens and enhancing resistance to biotic stresses like root rot and collar rot diseases. The symbiosis involves mechanisms such as creating a mechanical barrier, thickening cell walls through lignification, and producing polysaccharides

to hinder root pathogen entry. In drought stress, bioinoculants modify the root morphology and make the root denser to extend beyond the depletion zone, improving water uptake. Under salt stress, bioinoculants enhance resistance by accumulating solutes, improving osmotic adjustment, and increasing antioxidant production. In heavy metal toxicity, bioinoculants trap, immobilize, and accumulate metals, reducing their availability for plant uptake. Additionally, bioinoculants also contribute to secondary metabolite production, osmotic balance, photosynthesis, nitrogen fixation, and overall resilience to environmental and pathogenic stresses. Overall, bioinoculants significantly enhance nutrient and water uptake, reduce oxidative damage, and promote plant adaptation to stressful conditions.

4. **Increases fruit yield and quality:** Bio-inoculants boost plant yield by improving water and nutrient absorption, actively suppressing or eliminating pathogens, and fortifying plants against diseases, resulting in robust and healthy plants. It plays a crucial role in enhancing fruit quality by influencing morphological, physical, chemical, and organoleptic properties. By providing essential nutrients like phosphorus and zinc, AMF and PGPR contribute to the production of higher-quality fruits. Additionally, it produces growth hormones that play a vital role in elevating fruit quality. Inoculation with bio-inoculants significantly amplifies both plant growth and the presence of phytochemical constituents such as sugars, proteins, phenols, tannins, and flavonoids.

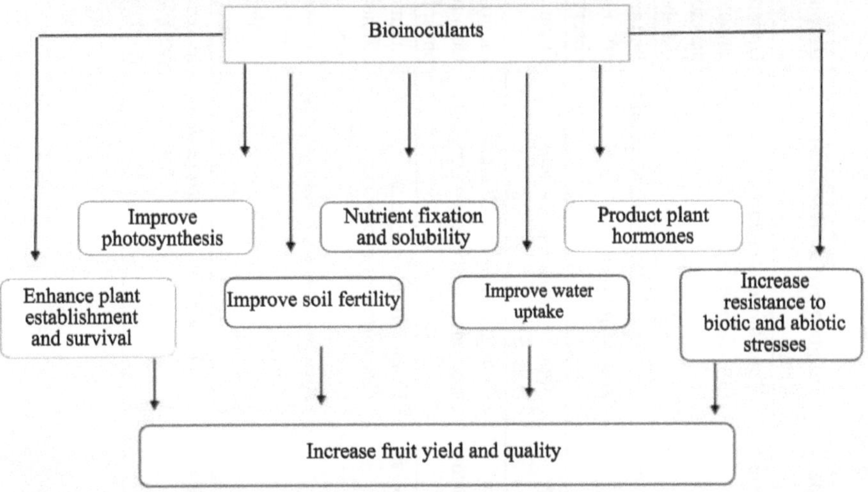

Fig. 3: Beneficial effects of bioinoculants

Table 1: Responses of horticulture plants to PGPR and VAM inoculation

Fruit plants	**Fungus Species/ Bacterial strain**	**Responses**	**References**
Trifoliate orange	*Funneliformis mosseae, Paragiomus occultum*	Under drought stress conditions, there was a significant enhancement in root biomass, lateral root growth, and the development of roots, along with increased sucrose and proline metabolisms.	Zhang *et al*. (2018).
Date palm	*Glomus sp., Sclerocystis sp.* and *Acaulospora sp.*	Improved relative water content, enhanced photosynthetic efficiency, and a strengthened antioxidant defense system were observed under salinity stress conditions.	Ait-El-Mokhtar *et al*. (2019)
Banana	*Bacillus amyloliquefaciens* NJN-6	Organic compound secretion and biofilm generation	Sen *et al*. (2015)
Kinnow Mandarin	*Glomus manihotis*, *Glomus mosseae*, and *Gigasporagigantia*	Improved growth parameters (plant height, canopy volume, leaf area, and number of new shoots per plant), increase plant phosphorous, potassium and calcium	Shamshiri *et al*. (2012)
Strawberry	*Funneliformis mosseae, Funneliformis geosporus* Pseudomonas sp, P. fluorescens	Augmented shoot and root fresh weights, improved water use efficiency (WUE), and increased plant survival percentage in the presence of drought stress. Enhanced anthocyanin levels in strawberries and decreased fertilizer usage. Elevated levels of sugar and anthocyanin, regulated pH, controlled malic acid, volatile compounds, and elemental concentrations.	Boyer *et al*. (2014) Lingua *et al*. (2013); Bona *et al*. (2015)
Pistachio	*G. mosseae* and *G. intraradices*	Increased concentrations of phosphorus (P), potassium (K), zinc (Zn), and manganese (Mn) were observed under both normal and drought stress conditions.	Bagheri *et al*. (2012)

Formulation and Application of Bio-inoculants

Various techniques create bio-inoculants with PGPR in solid or liquid forms, tailored for efficiency and desired CFUs. For soil fungi, root proximity suits, while aerial infections benefit from foliar application. Formulations vary by objective, land area, and timing. Though slower than agrochemicals, early use is advised. Synergy with antagonists improves effectiveness, even post-harvest. Applying on seeds or early plant growth boosts benefits. Bacteria like Pseudomonas or Bacillus foster growth and fight pathogens. VAM biofertilizer: 10g/plant at planting time is recommended.

Challenges in Bio-inoculant Application

Using bio-inoculants in fields for better crops and fighting plant diseases globally faces hurdles. These include short shelf life, and inconsistent results. Making them required specialized equipment, preventing cell death, and understanding microbe traits for field use. The consistent performance of bioinoculants, whether intended to promote plant growth or combat pests, remains a challenge. Maintaining the viability of bioinoculants poses difficulties; certain strains, such as *Bacillus sp.* with spore-forming abilities, have longer lifespans, while others require innovative methods to ensure their efficacy over time. Ensuring the presence of bioinoculants where plants require them most, such as in the root zone, is essential for their effectiveness. Applying them on plants in powder or liquid form is also tricky. Different places have different problems due to varied soil, weather, and places. Researchers are actively seeking local strains tailored to specific soils. Further investigation is necessary to understand how environmental factors impact the most effective use of each bioinoculant, leading to a deeper comprehension of their benefits.

Conclusion

In conclusion, sustainable agriculture strives for efficient crop growth while prioritizing social, economic, and environmental considerations. Emphasizing eco-friendly methods and minimizing harmful chemical use, sustainable horticulture integrates scientific research with practical approaches. Harnessing the power of beneficial microorganisms like PGPR and AMF is pivotal in revolutionizing horticultural practices. Drawing inspiration from ancient civilizations like the Mayans, who revered microorganisms for soil enrichment, underscores the ongoing evolution of these technologies. Preserving this ancestral wisdom holds promise for future generations, enhancing farming practices for a sustainable tomorrow.

References

Ait-El-Mokhtar, M., Laouane, R. B., Anli, M., Boutasknit, A., Wahbi, S., & Meddich, A., 2019. Use of mycorrhizal fungi in improving tolerance of the date palm (Phoenix dactylifera L.) seedlings to salt stress. Scientia Horticulturae, 253: 429-438.

Avio, L., Turrini, A., Giovannetti, M., & Sbrana, C., 2018. Designing the ideotype mycorrhizal symbionts for the production of healthy food. Frontiers in plant science, 9: 1089.

Bagheri, V., Shamshiri, M. H., Shirani, H. and Roosta, H., 2012. Nutrient uptake and distribution in mycorrhizal pistachio seedlings under drought stress. Journal of Agricultural Science and Technology, 14: 1591–1604.

Begum, N., Qin, C., Ahanger, M. A., Raza, S., Khan, M. I., Ashraf, M. & Zhang, L., 2019. Role of arbuscular mycorrhizal fungi in plant growth regulation: implications in abiotic stress tolerance. Frontiers in plant science, 10: 1068.

Bona, E., Lingua, G., Manassero, P., Cantamessa, S., Marsano, F., Todeschini, V., Copetta, A., D'Agostino, G., Massa, N., Avidano, L., Gamalero, E., & Berta, G., 2015. AM fungi and PGP pseudomonads increase flowering, fruit production and vitamin content in strawberry grown at low nitrogen and phosphorus levels. Mycorrhiza, 25(3):181-193.

Boyer, L. R., Brain, P., Xu, X. M., & Jeffries, P., 2015. Inoculation of drought-stressed strawberry with a mixed inoculum of two arbuscular mycorrhizal fungi: effects on population dynamics of fungal species in roots and consequential plant tolerance to water deficiency. Mycorrhiza, 25: 215-227.

Bücking, H., Liepold, E., & Ambilwade, P., 2012. The role of the mycorrhizal symbiosis in nutrient uptake of plants and the regulatory mechanisms underlying these transport processes. Plant Science, 4: 108-132.

Lingua, G., Bona, E., Manassero, P., Marsano, F., Todeschini, V., Cantamessa, S., Copetta, A., D'Agostino, G., Gamalero, E., & Berta, G., 2013. Arbuscular mycorrhizal fungi and plant growth-promoting pseudomonads increases anthocyanin concentration in strawberry fruits (Fragaria x ananassa var. Selva) in conditions of reduced fertilization. Int. International journal of molecular sciences, 14(8): 16207–16225.

Orozco-Mosqueda, M. D. C., Flores, A., Rojas-Sánchez, B., Urtis-Flores, C. A., Morales-Cedeño, L. R., Valencia-Marin, M. F., & Santoyo, G., 2021. Plant growth-promoting bacteria as bioinoculants: Attributes and challenges for sustainable crop improvement. Agronomy, 11(6): 1167.

Shamshiri, M. H., Usha, K., & Singh, B., 2012. Growth and Nutrient Uptake Responses of Kinnow to Vesicular Arbuscular Mycorrhizae. ISRN Agronomy, 2012, 1–7.

Shen, Z., Wang, B., Lv, N., Sun, Y., Jiang, X., Li, R. & Shen, Q., 2015. Effect of the combination of bio-organic fertilizer with Bacillus amyloliquefaciens NJN-6 on the control of banana Fusarium wilt disease, crop production and banana rhizosphere culturable microflora. Biocontrol science and technology, 25(6): 716-731.

Smith, S., & Read, D., 2008. Mycorrhizal symbiosis third edition introduction. Mycorrhizal Symbiosis, 1-9.

Wu, S., Zhang, X., Huang, L., & Chen, B., 2019. Arbuscular mycorrhiza and plant chromium tolerance. Soil Ecology Letters, 1: 94-104.

Zeng, R.S., 2006. Disease resistance in plants through mycorrhizal fungi induced allelochemicals. In Allelochemicals: biological control of plant pathogens and diseases, 181-192. Dordrecht: Springer Netherlands.

Zhang, F., Jia-Dong, H. E., Qiu-Dan, N. I., Qiang-Sheng, W. U., & Ying-Ning, Z. O. U., 2018. Enhancement of drought tolerance in trifoliate orange by mycorrhiza: changes in root sucrose and proline metabolisms. Notulae Botanicae Horti Agrobotanici Cluj-Napoca, 46(1): 270-276.

Zhu, B., Gao, T., Zhang, D., Ding, K., Li, C., & Ma, F., 2022. Functions of arbuscular mycorrhizal fungi in horticultural crops. Scientia Horticulturae, 303: 111-219.

5

Marketing of Feed and Fodder in India: Issues, Challenges and Way Forward

Bishwa Bhaskar Choudhary[1], Prem Chand[2], Priyanka Singh[3] and Sadhna Pandey[1]

[1]*ICAR-Indian Grassland and Fodder Research Institute, Jhansi, Uttar Pradesh*
[2]*ICAR-National Institute of Agricultural Economics and Policy Research New Delhi*
[3]*ICAR-Central Agroforestry Research Institute, Jhansi, Uttar Pradesh*

Abstract

The livestock sector has historically played a pivotal role in Indian agriculture, making substantial contributions to food, clothing, power, income, and employment. India with a significant livestock population of around 535 million and holding approximately one-fifth of the world's cattle, positioned itself as the world's largest milk producer. However, persistent concerns revolve around the suboptimal productivity of the Indian milch herd, attributed to intrinsic factors such as low genetic potential, and extrinsic factors like inadequate nutrition and feed management. Empirical studies underscore the critical impact of enhancing both the quality and quantity of feed and fodder to boost milk productivity. Currently, the nation grapples with a reported shortage of green fodder, dry fodder, and concentrates at 11.24%, 23.4%, and 28.9%, respectively, with certain states facing even more substantial deficits. Notably, feed and fodder alone constitute 60 to 70% of milk production costs. Addressing these challenges and optimizing the potential of the Indian dairy sector necessitates a comprehensive policy approach. This approach should include supporting fodder-based startups, ensuring a reliable market for surplus fodder, promoting silage business models, strengthening the extension network, and providing essential infrastructure support. Additionally, fostering awareness among farmer-entrepreneurs, tailoring investments for insurance mechanisms, and acknowledging commendable efforts in fodder development are indispensable steps toward achieving self-sufficiency in feed and fodder.

Keywords: *Livestock, Milk productivity, Fodder deficit, Fodder marketing*

Introduction

Livestock sector has been an integral component of Indian agriculture since time immemorial due to its multifarious contributions to the society in the form of nutrient-rich food products, clothing, drought power, income and employment. India has huge population of livestock. Of the total livestock population of the world, India alone has about one-fifth cattle population. As per the 20th livestock census estimate, total livestock population in the year 2019 was over 535 million. Among the livestock products, milk is the most important. India has a glorious history of outstanding achievement in transforming itself from milk deficit country to world's biggest milk producer over the span of three decades. With over 300 million bovines and producing over 198 million tonnes of milk in 2019-20, the Indian dairy sector exhibits strong growth potential. *Despite COVID-19 induced restrictions,* which has thrown the smallholder milk producers from the frying pan to fire, the organized dairy sector showed positive growth rate of around 1% in the last fiscal– the lowest in a decade. Despite these proud boasts, poor productivity potential of Indian milch herd is a major cause of concern that may threat its *numero one* position in future.

The major causes of low milk productivity in India are both intrinsic (low genetic potential) and extrinsic (poor nutrition/feed management). Empirical studies in India have shown that enhancing quality and quantity of feed and fodder has greater impact than breed improvement on increasing milk productivity (Lalwani, 1989; Gaddi and Kunal, 1996; IAEA, 2006; Roy *et al.*, 2020). The timely unavailability of nutritionally rich feed and fodder is a major hitch that impinges on the productivity growth of farm animals in the country. Nonetheless, with increase in agricultural production over the time, the animal feed availability has also improved, but its supply always falls short of the aggregate demand. At national level, the recent reported shortage in green fodder, dry fodder and concentrates is 11.24%, 23.4% and 28.9%, respectively (Roy *et al.* 2019), and the scenario is more unnerving in few Indian states where the fodder deficit is above 25% (Fig 1).

Further, with burgeoning livestock population and government focus on genetic upgradation of cattle by cross-breeding programmes, the demand-supply gap of feed and fodder will widen considerably, in absence of appropriate policy planning and its grass root level implementation.

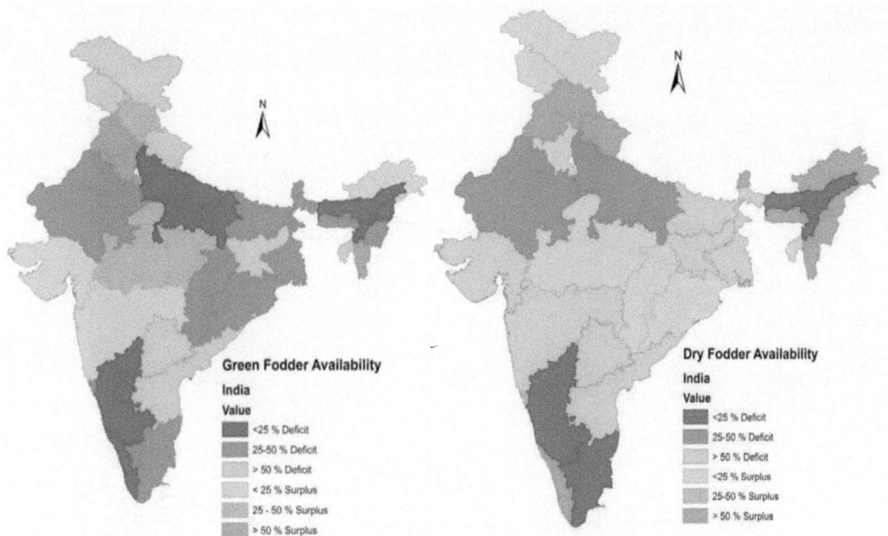

Source: Roy et al. (2019)

Fig. 1: Green and dry fodder availability scenario across Indian states

Inflationary Pressure of Feed and Fodder on Indian Dairy

The continuous rise in milk prices in the recent past has now attracted considerable policy attention as the unabated inflationary pressure has impacted the consumers who are already struggling with inflation across consumer product segment. A recent study by Local Circles indicated that, rising milk prices has led to one in three Indian households either reducing consumption or downgrading milk brand. Despite the presence of a strong cooperative in the India dairy sector, inflation has hit the sector like never before.

It is important to note that, there is no supply side constraints in milk reported yet, but the cost of production, procurement and transportation has gone up, leading to this price hike. Notably, feed and fodder alone accounts for 60 to 70 per cent of the production cost of milk (Sharma *et al.*, 2021; Choudhary *et al.*, 2022) and wholesale price index of cattle feed and fodder rose much faster than that of milk in the last decade.

Source: Ministry of Commerce and Industry, GoI

Fig. 2: Wholesale price index of cattle feed, fodder and milk (Base 2011-12)

In cattle feed production, about 90 per cent of the input is feed ingredients like, maize, brans, meals and oilcakes and feed additives like, multivitamins, mineral mixes, amino acids, etc. Presently, availability of feed ingredients is not an issue, but the sharp fluctuations in prices of the ingredients are a serious concern. The all-India trends in monthly wholesale price index of important feed ingredients clearly indicate the extent of seasonal fluctuations in prices. The WPI of cotton seed oil cake–one of the important feed used by dairy farmers in India – has remained higher than the WPI of milk and the gap between two is widening over the years (Fig. 2).

With the limited storage capacity available with the firms, especially the smaller firms, cost of production of feed is seriously affected. As a result of high volatility in prices, to maintain their profit margins, the firms often resort to substitution of feed ingredients at the expense of feed quality, thus impinging the productivity potential of livestock in long-run.

In fodder supplies, the paradoxical situation is that in many parts of the country there is surplus fodder during monsoon and a deficit occurring during lean season. This is especially so in the far flung areas. Few studies suggests that devoting around 14-17 per cent of land for fodder cultivation will be ideal for meeting fodder shortages in the country, but fodder is being cultivated on 8.4 mha (nearly 4%) area from last few decades of which nearly half is under cluster bean which is mainly used commercial crop rather than green fodder. Sparing more area for fodder is difficult in the wake of rapid intense competition for additional land from commercially important crops.

Existing Marketing System for Fodder Crops in India: A Brief Review

The market for dry and green fodders in India is highly informal. Farmers usually grow fodder crops in a small area and for feeding their livestock only. The surplus fodders, which the farmers normally have during normal monsoon seasons, are mostly sold within the village or nearby villages. As bulkiness nature of fodders tends to increase its transportation costs, hence inter area marketing of fodder takes place on very small scale. The marketing of fodder gets impetus only during summer season where fodder scarcity gets wide speared.

Available reviews suggest that existing marketing system in feed and fodders have been poorly explored in Indian context. Grover and Kumar (2012) in their consolidated report have highlighted marketing/processing systems for the disposal of fodder in the Indian states of Gujarat and Punjab, showing their size and efficiency. As per the report, in Gujarat, fodder is generally marketed through one marketing channel, namely Producer-Local Trader-Consumer. In this channel local trader incurred marketing expenses mainly on transportation and loading/unloading of fodder. The net profit margin of local trader on consumer's price was highest (14.9%) in rabi season followed by summer (9.2%) and lowest (8.9 %) in kharif season.

In Punjab, the following three major marketing channels were observed among sample fodder growers:

Channel-I: Producer-Forwarding agent/Commission agent-Dairy owner (Consumer)

Channel-II: Producer-Forwarding agent/Commission agent-Chaff cutter-Consumer

Channel-III: Producer –Consumer

In the *channel I*, the produce was directly taken by the producer to the forwarding/commission agent, who were forwarding the produce to the big dairy owners keeping in view the fodder demanded, through the chaff cutters. In *channel II*, the chaff cutter purchases the produce from forwarding/commission agent, who charges their commission from the producer as well as buyer. In *channel III*, the produce is directly disposed of to the consumers in the village itself. In *channel-I* for the sale of sorghum, the producer's share in consumer's rupee was found to vary from 74 to 77 per cent for the different fodder crops. In *channel-II* the producer's share in consumer's rupee was about 65 to 70 per cent for different crops. In a similar type of study in Karnataka, Kanan (2012) reported that except paddy straw, all the other fodders are disposed off by the farmers within the villages. Farmers in the state sale around 21 per cent of paddy straw outside the villages at price 10 per cent higher than the local market.

In an exploratory study from Bihar, Singh *et al.* (2013) highlighted the prevailing regional disparity in fodder availability in the state. South Bihar regions, due to irrigated cultivation of paddy and wheat, were found to be fodder surplus, while north Bihar was fodder deficit and hence depend on surplus regions for meeting fodder requirement. They further pointed out that, fodder trading usually takes place along roadsides and without legal credentials due to lack of dedicated market places for fodder. The most common fodder supply chain in the state begins with the producers and proceeds further along a number of different channels with the help of various kinds of actors such as assemblers and small vendors, commission agents, retailers, wholesalers and processors, and ends with the ultimate users who are scattered across the state (Fig 3).

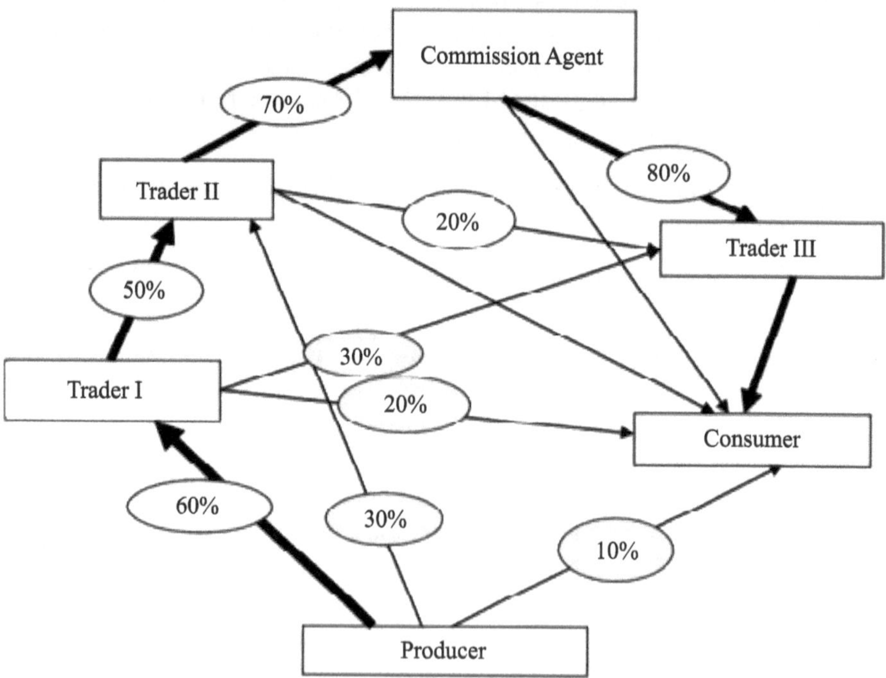

Source: Singh et al. (2013)

Fig. 3: Fodder transactions among different actors in Bihar

Processing of Fodder and Cost of Processing

Processing of fodders entails the conversion of raw fodder into a new form. Through processing, the fodder can be fed to animals as green feed. Processing is essentially done for two major reasons: (i) conservation and (ii) increase the palatability of feed. For increasing the palatability, fodders are chopped and mixed with some additives like salt and then fed to livestock instantly. But, the function of conservation/preservation of fodders helps the farmers to use them during lean seasons. The conservation function also helps to preserve the nutrients contents of the forages.

Generally, two methods of conservation of fodders are followed. They include ensilage and hay making. Ensilage is the process through which a material called silage is produced through anaerobic fermentation of crops. For making silage, container called silo is used. Silos are of different types like pit silo, drench silo and drum silo (Sharma and Choudhary, 2021). Drying of forages is called hay making. Forages are generally dried under sunlight. The harvested forages or cereals straw are spread on the ground under sunlight and layers are changed regularly for drying evenly. The dried forages are staked in heaps for use. As compared to hay making, ensilage produces high quality feed with rich in nutrients. But, lot of care is required to make better silage under controlled conditions; otherwise it will result in bad quality and wastage.

In a recent study conducted by Singh *et al.* (2018) in the state of Punjab indicated that making silage pits (1m^3) costs on an average Rs. 17,904. Cost involved in silage making from various fodder crops of Punjab state is presented in Table 1. A perusal of the table revealed that the farmers involved in the processing of green fodder were practicing silage making for three crops namely maize, bajra, and sorghum. A scrutiny of table brought out that the variable cost for maize crop which includes loading/unloading, transportation, chaffing, etc. was the highest i.e. Rs. 20.20/q out of which harvesting cost was maximum i.e. Rs. 4.91/q followed by the transportation cost Rs. 4.79/q and plastic sheet used Rs. 3.45/q. Similarly, the variable cost of silage making was Rs. 18.77/q and Rs. 14.70/ q for bajra and sorghum crops, respectively. The total cost of silage making was the highest in maize crop with Rs. 31.50/q followed by the sorghum (Rs. 30.98/q) and bajra (Rs. 30.07/q). The processing of fodder in the form of silage saves the labour cost and helps in reducing the variable cost of milk production by about 8-10 per cent.

Further, regarding the other parameters of silage making, the amount stored in pit varied between 1300 and 2100 quintals, period of storage from 6 to 10 months and period required for preparation of silage from 43-47 days for different crops. About 76.92 percent of maize was home grown and 23.08

percent of it was being purchased from outside. Similarly, for bajra, 88.90 percent was home grown fodder and rest 11.11 percent was purchased from outside. For sorghum crop, there was no purchase from outside as whole crop was home grown. There was approximately 3-4 percent loss in preparation of silage.

Table 1: Post-harvest operational cost (Rs./q) in Silage making from various fodder crops in Punjab state, 2017-18

Operational costs	Maize	Bajra	Sorghum
Harvesting	4.91	4.68	2.93
Loading/unloading	2.85	3.13	3.07
Transportation	4.79	4.47	5.42
Chaffing	2.13	2.03	1.64
Chemical	2.08	1.38	
Plastic sheet	3.45	3.12	1.64
Total	20.21	18.77	14.70
Other parameters of silage making			
Average amount stored (q)	2068.84	1563.33	1295
Period of storage (months)	8	7	9-10
Period required for preparation (days)	43	47	45
Home grown fodder (%)	76.92	88.90	100.00
Percent of silage prepared	96.82	96.89	97.65

Source: Singh *et al.*, (2018)

Hay making is easier, but it is difficult during rainy season. The quality of hay making is influenced by the seasonal conditions like forage availability and weather. Fodder cereals like jowar, maize and bajra are cultivated mainly during kharif seasons; however in southern states like Karnataka, it is cultivated both in kharif and rabi seasons. Crop residues available during these seasons are made use of by the famers for hay making.

In a study conducted by Agro-Economic Research Centre of Punjab Agricultural University, Ludhiana indicated that in the Indian state of Gujarat (Grover *et al.*, 2012), many farmers prefers to make hay during kharif season as bright sunshine during summer enables the farmers to make good quality hay. All farmers were found using plastic/tarpaulin sheet to cover hay. This practice saves the hay from development of moulds. Chemical like, BHC/Gamxene/Phorate were used by one fourth of hay making households to prevent the damage caused by insects and pests.

Many farmers said that they avoided use of chemical as it changes the smell and taste of fodder and the bovines do not prefer this kind of smell of fodder. The fodder was stored maximum for 140 days and minimum for 46 days. The storage cost per Qtl. per month range from Rs. 2.30 to Rs. 3.30 in kharif and

Rs. 3.00 to 3.10 in rabi and Rs. 2.90 to Rs. 3.40 in summer. Average quantity stored per household was 57.91 Qtl. in kharif and 100.93 Qtl. in rabi (Table 2). Loss of produce during the storage period was around 14 %. Among various operational costs, share of harvesting in total cost was maximum. The other major cost items were transportation, loading/unloading and storage.

Table 2: Post-harvest operational cost (Rs./q) in Hay making from various fodder crops in Gujarat

Operational costs	**Kharif**	**Rabi**
Harvesting	9.28	5.65
Packing	3.21	2.61
Loading/unloading	2.79	4.78
Transportation	3.98	5.22
Chaffing	1.18	-
Storage	2.65	2.61
Chemical	0.74	0.87
Other	1.67	1.74
Total	25.50	23.48
Other parameters of Hay making		
Average quantity of produce stored	57.91	100.63
Material used for storage (%) Sheet Chemical	 100 24.62	 100 37.50
Produce lost during storage (%)	14.14	14.18

Source: Grover *et al.*, (2012)

Feed block making could be good strategies for reducing the cost involved in transportation of fodder from one place to other and save the storage space. The mechanization aspect may also be thought of in terms of harvesting with weed cutters and chaffing of fodder with power operated chaff cutters, which reduce the reliance on manual labour and also help in saving time on these activities. It will also help in supplying fodder during the calamities as well as lean season. Complete feed block can be prepared by mixing dry fodder and concentrate mixture in certain proportion that meets the animal requirement. The roughage concentrate ration may be 70:30 to 40:60. The important consideration is that the CP and TDN content in feed block should be more than 12% and 60%, respectively (ICAR-IGFRI, 2022).

Quality Enhancement of Fodder

Wheat and paddy straw and stover of millets (maize, pearl millet and sorghum) are major source of dry fodder. Urea treatment of straw increases its nitrogen content resulting into enhanced microbial activity and ruminal digestion of the straw. In addition, urea treatment also exerts its effect on lingo-cellulose complex, wherein the lignin forms the complex with cellulose, thus preventing its microbial digestion. Urea also acts as preservative and application of 3-4% urea solution on the straw and subsequent air tight storage of treated straw for 14-21 days, depending on the ambient temperature, if it is less than 20^{0}C, then 21 days air tight storage would ensure the proper treatment of straw. At higher ambient temperature 14 days storage is sufficient for straw treatment. The technology although is very old but it still hold promise to improve the feed basket of the livestock. The use of a cheap source of nitrogen such as urea to improve the nitrogen content of such roughages makes a promising alternative to improve the nutritive value of straw. Further spray of salt and mineral mixtures also enhances the palatability and nutritive value of dry fodders.

SWOT Analysis of Feed and Fodder Marketing in India and Policy Issues

Various state research institutions along with the IGFRI, Jhansi – a national research institute under the administrative control of Indian Council of Agricultural Research, New Delhi has developed a number of improved fodder crop varieties and technologies that can ensure year round availability of quality feed and fodder. These fodder crops and grasses are less input intensive than the field crops. However, knowledge and adoption of improved fodder varieties in India has remained low. Though constraints in adopting improved fodder technologies by farmers vary locally, lack of assured market for fodder is unanimous challenge hindering in maintaining fodder balance for the country as whole.

Therefore an urgent policy need is to ensure parallel development of supporting market environment for surplus fodder encompassing backward and forward market linkages. Besides, a sincere focus on 4-S strategies (i) Smoothening credit facility for forage production (ii) Support price for forage and marketing of seed (iii) Silage business model involving seed firms, operations service providers (for baling and supply chain functions) and rural retail channels and (iv) Strengthening extension network would pave a long way in achieving the full potential of the Indian dairy herd in long run.

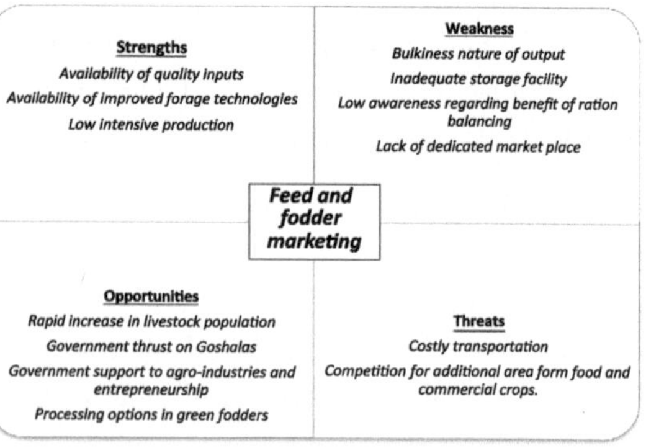

The vision of becoming "atmanirbhar" in feed and fodder will not be materialized without proactive role of both central and state governments in supporting fodder based start-ups. Government should give buy back guarantee with prior agreement for procurement of certified fodder seed and surplus fodder from the entrepreneurs. Infrastructure facilities like silage and hay balers, fodder block making units, chaff cutter, crop harvesters, pivot irrigation equipment etc along with all facilities required should be supported. Construction of more fodder banks at block level of all fodder stress districts must be a prior policy focus.

Government has to support fodder cultivation as bankable project to those entrepreneurs who wish to take up fodder cultivation as an economic activity to obtain bank loans. Awareness need to generated among farmer-entrepreneurs for making value additions to the fodder to use it as a nutritious food for animals and thereby saving a large amount of fodder that goes waste every year. Moreover, investment must be tailored for developing insurance mechanism for mitigating the financial risks faced by fodder based start-ups. Lastly, government must recognise entrepreneurs as well as concerned institutions and organizations which have done commendable work in the field of fodder development by way of merit certificates and cash incentives.

Conclusion

The present chapter focuses on how inadequate availability of nutritionally rich feed and fodder hinders the productivity potential of Indian milch herd and thus the milk productivity. Also the sharp hike in fodder inflation played a major role in rising milk prices as feed and fodder alone accounts for 60 to 70 per cent of the production cost of milk. Therefore, efforts must be directed towards developing *"fodder banks"* to store surplus fodder during monsoon and meeting the deficits during lean season. There is a compelling need for concerted efforts from government to bring additional area under fodder cultivation to address the fodder deficit issues. Strengthening market & value chains and partnership in fodder trade with other countries would be imperative in meeting regional and seasonal disparity in fodder and quality feed for livestock. In light of this, Public-Private partnership in fodder sector needs to be promoted to take up production of quality compound feed to cater to the nutritional requirements of animals of different productive potential. Though fodder-based research organizations have come up with various innovative technologies to offer quality feeding solutions to dairy farmers, their ground level adoption is very limited across the country. Imparting training and capacity building of potential youth farmers and providing them efficient resources (physical, technical and networking support) would be crucial for successful establishment of enterprises and agripreneurs in fodder sector. Furthermore, the fodder crop should also get central place within the agricultural system and bestowed with all the perks of agricultural crops like crop insurance, minimum support price and other benefits. Lastly, fostering a culture of collaboration and knowledge sharing among stakeholders will be pivotal in ensuring year round availability of fodder crops across the country.

References

Choudhary, B.B., Sharma, P., Singh, P., Kumar, S., Gupta, G., Kantwa, S.R., Upadhyay, D., Wasnik, V.K., Prasad, M. and Sharma, R.K. 2022. Green fodder cultivation improves technical efficiency of dairy farmers in semi-arid tropics of central India: a micro-analysis. Journal of Dairy Research, 89: 367–374. https://doi.org/ 10.1017/S0022029922000759.

Gaddi, G.M. and Kunnal, L.B. 1996. Source of output growth in new milk production technology and some implications to return on research investment. Indian Journal of Agricultural Economics, 51(3): 389-395.

Grover, D.K. and Kumar, S. 2012. Economics of production, processing and marketing of fodder crops in India. AERC Study No. 29, Agro-Economic Research Centre, Department of Economics and Sociology Punjab Agricultural University Ludhiana.

IAEA. 2006. Improving animal productivity by supplementary feeding of multi-nutrient blocks, controlling internal parasites, and enhancing utilization of alternate feed resources. Publication prepared under the framework of an RCA project with technical support of the Joint FAO/IAEA Programme of Nuclear Techniques in Food and Agriculture.

ICAR-IGFRI. 2022. Fodder resource development plan for Punjab. ICAR-Indian Grassland and Fodder Research Institute, Jhansi.

Kanan, E. 2012. Economics of production, processing and marketing of fodder crops in Karnataka. Research Report: IX/ADRTC/142, Agricultural Development and Rural Transformation Centre Institute for Social and Economic Change Bangalore - 560 072.

Lalwani, M. 1989. Technological change in India's dairy farming sector, distribution and output gains. Indian Journal of Agricultural Economics, 44(1): 55-66.

Roy, B.C., Mondal, B., Majumder, D., Biswas, R.K. and Roy, A. 2020. Assessment of livestock feed and fodder in the state of West Bengal. Study No.-191, Agro-Economic Research Centre (For the States of West Bengal, Sikkim and Andaman & Nicobar Islands), Visva-Bharati, Santiniketan, West Bengal.

Sharma, P., Choudhary, B.B., Singh, P., Kumar, S., Gupta, G. and Dev, I. 2021. Can forage technologies transform Indian livestock sector? Evidences from smallholder dairy farmers in Bundelkhand region of central India. Agricultural Economics Research Review, 34: 73–82.

Sharma, P. and Choudhary, B. B. 2021. Planning for Dairy Farm. In: P Sharma., S Phand., B.B. Choudhary, G. Gaurednra, R.K. Sharma (eds.) Agripreneurship Development on Value Added Fodder Products [E-book]. Hyderabad: National Institute of Agricultural Extension Management & ICAR-Indian Grassland and Fodder Research Institute, Jhansi (U.P.), 118-139.

Singh, H., Singh, V.P. and Kaur, I. 2018. Cost analysis of processing of green fodder crops in Punjab state. International Journal of Current Microbiology and Applied Sciences, 7(11): 832-837.

Singh, K.M., Singh, R.K.P., Jha, A.K. and Kumar, A. 2013. Fodder market in Bihar: an exploratory study. MPRA paper no. 53597.

6

Value of Soil Health in Agriculture

Hari Shankar Singh[1], Mayank Kumar[2] and Sumit Kumar[3]

[1,2,3]Department of Soil Science and Agricultural Chemistry, CoA, CSAUAT Kanpur, Uttar Pradesh

Abstract

The health of soil is a cornerstone of agricultural sustainability and productivity, with far-reaching implications for global food security, environmental conservation, and human well-being. In this comprehensive exploration, we underscore the critical importance of prioritizing soil health in agricultural practices. By examining various soil health indicators and management strategies, we elucidate the multifaceted value of soil health in sustaining agricultural systems and mitigating environmental degradation. Soil health encompasses a complex interplay of physical, chemical, and biological properties, all of which influence crop productivity, nutrient cycling, and ecosystem resilience. Physical properties, including soil texture, structure, and porosity, govern water infiltration, root penetration, and soil aeration, crucial factors for plant growth and soil stability. Chemical properties, such as nutrient content, pH, and cation exchange capacity (CEC), determine nutrient availability and soil fertility, essential for sustaining crop yields and maintaining soil productivity. Biological components, including microbial biomass, diversity, and activity, play pivotal roles in nutrient cycling, organic matter decomposition, and soil ecosystem functioning, contributing to overall soil health and ecosystem resilience. Adopting sustainable soil management practices is paramount to safeguarding soil health and ensuring long-term agricultural sustainability. Conservation tillage, cover cropping, crop rotation, and agro-forestry are among the practices that promote soil conservation, improve soil structure, enhance organic matter content, and mitigate erosion and nutrient runoff. Integrating these practices into agricultural systems not only enhances soil health but also contributes to climate change mitigation, water conservation, and biodiversity conservation. Empirical evidence, and practical insights, we illustrate the economic, ecological, and societal benefits of investing in soil health initiatives. By integrating soil health considerations into agricultural policies and practices, we can ensure the resilience and sustainability of farming systems while safeguarding natural resources and mitigating climate change impacts. Ultimately, prioritizing soil health is essential for securing the future of food production and environmental sustainability in a rapidly changing world.

Keywords: *Soil Health, Soil Productivity, Soil Management, Yield, Efficiency, etc.*

Introduction

Soil, the complex mixture of minerals, organic matter, water, air, and living organisms, forms the foundation of terrestrial ecosystems and sustains life on Earth. Its significance in agriculture transcends mere physical support for plant roots; soil health is intricately linked to crop productivity, environmental sustainability, and human well-being. As navigate the challenges of feeding a growing global population, understanding and prioritizing soil health in agricultural practices becomes imperative. This gateway to explore the multifaceted value of soil health in agriculture. We begin by elucidating the critical role soil plays in supporting agricultural productivity and food security. The various components of soil health and their interrelationships, emphasizing the dynamic nature of soil ecosystems. Furthermore, we highlight the importance of adopting sustainable soil management practices to mitigate environmental degradation and enhance agricultural resilience. Throughout this discussion, we draw upon a diverse array of research findings and scholarly literature to underscore the significance of soil health as a cornerstone of sustainable agriculture.

Soil health plays a critical role in agricultural productivity, sustainability, and ecosystem resilience. Healthy soil provides essential nutrients, supports diverse microbial communities, improves water retention, and enhances crop resilience to pests and diseases. As stated by the United Nations Food and Agriculture Organization (FAO), maintaining soil health is vital for ensuring global food security and mitigating the impacts of climate change. Numerous studies have emphasized the economic benefits of investing in soil health practices, such as cover cropping, crop rotation, and reduced tillage, which can improve yields, reduce input costs, and enhance long-term soil productivity. Therefore, prioritizing soil health management strategies is essential for sustainable agriculture and securing food production for future generations.

Soil is the sustaining factor for life on earth. The upper crust of the Earth is full of organisms, liquids, and mineral particles that plants need to grow and thrive. Plants rely on the soil for everything from disease resistance to winter hardiness to fruit quality and flavor. Everyone who eats benefits from the quality of soil. 95% of food production is soil-based! As the population grows, there are more mouths to feed, straining soil as a natural resource. Even today, over 2 billion people experience micronutrient deficiencies in their diet, which impact factors of human health ranging from brain and muscle function to fetal development and immune system health.

The Role of Soil in Agricultural Productivity and Food Security

At its core, agriculture relies on the fertility and productivity of soil to sustain crop growth and ensure food security for billions of people worldwide. Soil serves as a reservoir for essential nutrients, water, and air-key elements that plants require for healthy development. Moreover, soil acts as a habitat for a myriad of organisms, including bacteria, fungi, earthworms, and insects, which play crucial roles in nutrient cycling, pest control, and soil structure maintenance. Soil plays a crucial role in agriculture productivity and food security by providing essential nutrients, anchoring plants, and regulating water and air flow. Healthy soil promotes better crop growth, higher yields, and resilience to pests and diseases. Sustainable soil management practices are vital for maintaining long-term agricultural productivity and ensuring food security for the growing global population.

1. **Nutrient Provision:** Soil serves as a reservoir of essential nutrients such as nitrogen, phosphorus, potassium, and micronutrients like zinc and iron. These nutrients are vital for plant growth, development, and overall health. Soil organic matter, derived from decomposed plant and animal residues, also contributes to nutrient availability.

2. **Physical Support:** Soil provides the physical support necessary for plant root systems to anchor and establish themselves. This anchorage enables plants to access water and nutrients from the soil, as well as withstand external stresses such as wind and rain.

3. **Water Regulation:** Soil plays a critical role in regulating water availability for plants. It acts as a sponge, capable of storing and releasing water as needed by plants. Proper soil structure and organic matter content improve water infiltration and retention, reducing the risk of waterlogging or drought stress.

4. **Air Exchange:** Soil provides a medium for gas exchange, allowing oxygen to reach plant roots and facilitating the diffusion of carbon dioxide produced during root respiration. Adequate soil aeration is essential for root growth and microbial activity, which contribute to nutrient cycling and soil fertility.

5. **Biological Diversity:** Soil is teeming with a diverse array of microorganisms, including bacteria, fungi, protozoa, and earthworms, among others. These organisms play vital roles in nutrient cycling, decomposition of organic matter, and suppression of plant pathogens, thereby enhancing soil fertility and plant health.
6. **Carbon Sequestration:** Healthy soils act as a significant carbon sink, sequestering carbon dioxide from the atmosphere through the process of photosynthesis and storing it in organic matter and soil minerals. Carbon sequestration in soils helps mitigate climate change by reducing greenhouse gas emissions and improving soil quality.
7. **Erosion Control:** Soil erosion, caused by water and wind, poses a significant threat to agricultural productivity and food security. Maintaining soil cover through vegetation, conservation tillage practices, and erosion control structures helps prevent soil loss, safeguarding the integrity and fertility of agricultural land.
8. **Sustainable Management:** Adopting sustainable soil management practices such as crop rotation, cover cropping, organic amendments, and reduced tillage promotes soil health and resilience. These practices enhance soil structure, fertility, and biological activity while minimizing environmental impacts and preserving natural resources for future generations.

Components of Soil Health and Their Interrelationships

Soil health is a holistic concept that encompasses various physical, chemical, and biological properties of soil. Physical properties such as texture, structure, and porosity influence water infiltration, root penetration, and soil aeration, thereby affecting plant growth and soil resilience. Chemical properties, including nutrient content, pH, and Cation Exchange Capacity (CEC), determine nutrient availability and soil fertility. Biological components, such as microbial biomass, diversity, and activity, are vital for nutrient cycling, organic matter decomposition, and soil ecosystem functioning. Soil health is determined by the interactions among various physical, chemical, and biological components within the soil ecosystem. Here's a detailed breakdown of the key components of soil health and their interrelationships:

1. Physical Components

Soil Texture: Soil texture refers to the relative proportions of sand, silt, and clay particles in the soil. It influences water retention, drainage, and aeration. Sandy soils have larger particles and drain quickly but may have poor water and nutrient retention. Clay soils have smaller particles and hold more water and nutrients but can become compacted.

Soil Structure: Soil structure refers to the arrangement of soil particles into aggregates or clumps. Well-aggregated soils have good porosity, allowing for air, water, and root movement. Soil aggregation is influenced by organic matter content, microbial activity, and management practices such as tillage.

Soil Porosity: Porosity refers to the volume of open space or pores in the soil. It affects water infiltration, air exchange, and root growth. Soil porosity is influenced by soil texture, structure, compaction, and organic matter content. Proper soil porosity is essential for maintaining soil moisture and oxygen levels for plant growth and microbial activity.

2. Chemical Components

Soil pH: Soil pH is a measure of soil acidity or alkalinity on a scale from 0 to 14. It influences nutrient availability, microbial activity, and soil structure. Most crops prefer slightly acidic to neutral soils (pH 6-7). Soil pH can be modified through liming or acidification to optimize nutrient availability and crop performance.

Nutrient Availability: Soil nutrients, including nitrogen, phosphorus, potassium, and micronutrients, are essential for plant growth and development. Nutrient availability is influenced by factors such as soil pH, organic matter content, mineralization, and nutrient cycling processes mediated by soil microbes. Balanced nutrient management is crucial for maintaining soil fertility and supporting healthy crop growth.

Cation Exchange Capacity (CEC): CEC is a measure of soil's ability to retain and exchange positively charged ions (cations) such as calcium, magnesium, potassium, and ammonium. It is determined by soil texture, organic matter content, and clay mineralogy. Soils with higher CEC have greater nutrient retention capacity and are more resilient to nutrient leaching.

3. Biological Components

Soil Organic Matter (SOM): SOM consists of decomposed plant and animal residues, microbial biomass, and humus. It serves as a source of nutrients, energy, and habitat for soil organisms. SOM improves soil structure, water retention, nutrient cycling, and biological activity. Management practices that increase SOM, such as cover cropping, composting, and reduced tillage, enhance soil health and productivity.

Soil Microorganisms: Soil microbes, including bacteria, fungi, protozoa, and nematodes, play critical roles in nutrient cycling, organic matter decomposition, and disease suppression. Beneficial microbes form symbiotic relationships with plants, enhancing nutrient uptake and resilience to stress. Soil microbial diversity and activity are influenced by factors such as soil moisture, temperature, pH, and organic matter availability.

Soil Fauna: Soil fauna such as earthworms, arthropods, and nematodes contribute to soil health by decomposing organic matter, improving soil structure through burrowing, and regulating microbial populations. Earthworms, for example, enhance soil aeration and nutrient cycling by consuming organic residues and excreting nutrient-rich casts.

4. Interrelationships

Physical, chemical, and biological components of soil health are interconnected and influence each other's dynamics. For example, soil structure influences water infiltration, which affects nutrient availability and microbial activity. Organic matter content influences soil aggregation, nutrient cycling, and microbial diversity. Soil pH affects microbial community composition, nutrient availability, and soil structure. Management practices such as cover cropping, crop rotation, reduced tillage, and organic amendments can improve multiple aspects of soil health simultaneously by enhancing soil structure, increasing organic matter content, balancing nutrient availability, and promoting beneficial microbial activity. Healthy soils with balanced physical, chemical, and biological properties exhibit greater resilience to environmental stressors such as drought, disease, and erosion. Sustainable soil management practices that enhance soil health contribute to long-term agricultural productivity, environmental sustainability, and food security.

Adopting Sustainable Soil Management Practices

In light of mounting environmental challenges, such as soil erosion, nutrient depletion, and pollution, adopting sustainable soil management practices is paramount to safeguarding soil health and ensuring long-term agricultural sustainability. Practices such as conservation tillage, cover cropping, crop rotation, and agro-forestry promote soil conservation, improve soil structure, enhance organic matter content, and mitigate erosion and nutrient runoff. Adopting sustainable soil management practices is crucial for maintaining soil health, enhancing agricultural productivity, and ensuring long-term food security. Here are some key sustainable soil management practices:

1. **Conservation Tillage:** Reduce or eliminate tillage to minimize soil disturbance and preserve soil structure. Conservation tillage practices such as no-till or reduced tillage help prevent soil erosion, improve water infiltration, and promote soil organic matter accumulation.
2. **Cover Cropping:** Plant cover crops, such as legumes or grasses, during fallow periods to protect soil from erosion, suppress weeds, and add organic matter. Cover crops also enhance soil fertility by fixing nitrogen and improving soil structure through root growth. Crop Rotation: Rotate different crops in sequence to diversify plant species, break pest and disease cycles, and improve soil nutrient balance. Crop rotation helps maintain soil fertility, reduce the buildup of pests and pathogens, and enhance soil microbial diversity.
3. **Organic Amendments:** Incorporate organic materials such as compost, manure, or green manure into the soil to increase soil organic matter content, improve nutrient availability, and enhance soil structure. Organic amendments also contribute to carbon sequestration and reduce reliance on synthetic fertilizers.
4. **Integrated Nutrient Management:** Optimize nutrient use efficiency by integrating organic and mineral fertilizers based on soil testing and crop nutrient requirements. Balanced nutrient management practices minimize nutrient runoff and leaching, reduce environmental pollution, and promote soil fertility.
5. **Conservation Practices:** Implement erosion control measures such as contour farming, terracing, and vegetative buffers to prevent soil erosion by water and wind. Conservation practices protect soil from degradation, preserve soil moisture, and maintain soil productivity.
6. **Agroforestry and Alley Cropping:** Integrate trees or shrubs with agricultural crops to enhance soil fertility, provide shade and wind protection, and diversify farm income. Agroforestry systems improve soil structure, increase biodiversity, and offer multiple ecosystem services.
7. **Water Management:** Implement irrigation and drainage systems to optimize water use efficiency and prevent waterlogging or drought stress. Proper water management practices maintain soil moisture levels, support crop growth, and minimize soil erosion.
8. **Soil Testing and Monitoring:** Regularly test soil nutrient levels, pH, and organic matter content to guide nutrient management decisions and

assess soil health. Soil monitoring helps identify potential issues early and allows for timely adjustments to management practices.

9. **Education and Outreach:** Provide farmers with training, technical assistance, and access to information on sustainable soil management practices. Extension services, farmer field schools, and demonstration plots can promote knowledge sharing and facilitate adoption of sustainable soil management techniques.

Increased Crop Yield and Quality

- Healthy soil promotes better root development and nutrient uptake, leading to increased crop yields.
- Higher quality crops fetch premium prices in the market, resulting in increased economic returns for farmers.

Reduced Input Costs

- Healthy soil requires fewer inputs such as fertilizers and pesticides, reducing production costs for farmers.
- Lower input costs contribute to improved profit margins and overall economic viability of agricultural operations.

Enhanced Soil Structure and Stability

- Soil health practices like cover cropping and reduced tillage improve soil structure, reducing soil erosion and loss of valuable top soil.
- Stable soil structure reduces the need for costly soil erosion control measures, saving farmers money in the long run.

Improved Water Management

- Healthy soil has better water infiltration and retention capacity, reducing irrigation needs and conserving water resources.
- Efficient water management minimizes water-related expenses and ensures more consistent crop yields, contributing to economic stability for farmers.

Long-Term Sustainability and Resilience

- Investing in soil health practices builds resilience against environmental stresses such as droughts and floods, reducing crop losses and economic risks.

- Sustainable farming practices contribute to the long-term viability of agricultural operations, safeguarding farmers' income and livelihoods.

Market Access and Consumer Demand

- Consumers increasingly value sustainably produced food, driving demand for products grown using soil health practices.
- Access to premium markets and certification programs for sustainably produced goods can command higher prices and increase farm income.

Regulatory Compliance and Risk Mitigation

- Compliance with soil health regulations and conservation programs may provide farmers with financial incentives or subsidies.
- Implementing soil health practices reduces the risk of regulatory penalties and environmental liabilities, protecting farmers' financial interests.

In the subsequent sections of this chapter, we delve deeper into each of these themes, elucidating the economic, ecological, and societal dimensions of soil health in agriculture. Through case studies, empirical evidence, and practical insights, we aim to underscore the imperative of prioritizing soil health considerations in agricultural policies and practices to secure the future of food production and environmental sustainability.

Conclusion

The value of soil health in agriculture cannot be overstated. Throughout this chapter, we have examined the myriad ways in which soil health influences agricultural productivity, environmental sustainability, and human well-being. From its role as a vital medium for plant growth to its function as a reservoir for essential nutrients and a habitat for diverse microbial communities, soil health underpins the foundation of agricultural systems worldwide. We have highlighted the importance of adopting sustainable soil management practices to safeguard soil health and ensure the long-term viability of agricultural production. Practices such as conservation tillage, cover cropping, crop rotation, and agro-forestry not only improve soil structure and fertility but also mitigate erosion, nutrient runoff, and greenhouse gas emissions. By integrating these practices into agricultural systems, farmers can enhance soil health, conserve natural resources, and mitigate the adverse impacts of climate change on agricultural productivity. Furthermore, we have explored the potential of emerging technologies and innovations to revolutionize soil health assessment and management. Precision agriculture, remote sensing, and soil microbiome research offer promising avenues for monitoring soil conditions, optimizing input use, and enhancing soil health and agricultural productivity. By harnessing the power of these technologies, farmers can make informed decisions to improve soil health and enhance the resilience and sustainability of farming systems. Overall, prioritizing soil health is essential for securing the future of food production and environmental sustainability in a rapidly changing world. By investing in soil health initiatives and integrating soil health considerations into agricultural policies and practices, we can ensure the resilience and sustainability of agricultural systems while safeguarding natural resources and mitigating climate change impacts. Ultimately, the value of soil health extends beyond agricultural productivity; it is essential for sustaining life on Earth and ensuring the well-being of future generations.

References

Doran, J. W., & Zeiss, M. R. 2000. Soil health and sustainability: managing the biotic component of soil quality. Applied Soil Ecology, 15(1), 3-11.

FAO. 2020. Status of the World's Soil Resources: Main Report. Food and Agriculture Organization of the United Nations.

Gupta, R., Gowda, J. H., Bhattacharyya, R., & Srivastava, S. 2021. Impact of conservation agriculture on soil health, crop productivity, and farmer's profitability: A review. Soil Systems, 5(2), 34.

Kibblewhite, M. G., Ritz, K. and Swift, M. J. 2008. Soil health in agricultural systems. Philosophical Transactions of the Royal Society B: Biological Sciences, 363(1492), 685-701.

Lal, R. 2015. Soil health and carbon management. Food and Energy Security, 4(3), 197-207.

Lal, R. 2018. Digging deeper: A holistic perspective of factors affecting soil organic carbon sequestration in agroecosystems. Global Change Biology, 24(8), 3285-3301.

M. Tahat, M.; M. Alananbeh. *et al.*, 2020. Soil Health and Sustainable Agriculture. Sustainability, 12.

Pimentel, D. and Burgess, M. 2013. Soil erosion threatens food production. Agriculture, Ecosystems & Environment, 46(1), 9-20.

Schipanski, M. E., MacDonald, G. K., Rosenzweig, S., Chappell, M. J., Bennett, E. M., Kerr, R. B., & Crews, T. 2016. Realizing resilient food systems. BioScience, 66(7), 600-610.

Six, J., Bossuyt, H., Degryze, S., & Denef, K. 2004. A history of research on the link between (micro) aggregates, soil biota, and soil organic matter dynamics. Soil and Tillage Research, 79(1), 7-31.

7

Preservation Pioneers: Understanding the Fundamentals and Techniques of Food Preservation

Deeksha Semwal[1] and Purnita Raturi[*2]

[1]Department of Food Science & Technology, CoA, GBPUAT, Pantnagar Uttarakhand

[2]Department of Horticulture, CoA, GBPUAT, Pantnagar, Uttarakhand

Abstract

Food preservation refers to the process or techniques employed to maintain specific desirable properties and quality of food in order to derive maximum benefits. An effective method of preserving food is one that hinders or completely prevents the activities of spoilage agents without causing harm to the food itself. The choice of preservation method depends on the type of food being preserved. Throughout human history, food preservation has played a crucial role due to the seasonal fluctuations in the availability of various foods. Preservation allows for the consumption of certain foods during periods of scarcity, ensuring a year-round food supply. Typically, food preservation involves managing or inhibiting the growth of microorganisms and minimizing quality deterioration caused by microbial spoilage or undesired chemical changes, such as the oxidation of fats leading to rancidity. Modern food preservation has evolved into a complex, interdisciplinary field of science. Recent years have seen the development of advanced preservation techniques aimed at extending shelf life, reducing risks, safeguarding the environment, and enhancing the functional, sensory, and nutritional aspects of preserved foods. Numerous sophisticated preservation technologies have been introduced, some of which are already in commercial use for specific applications, while others show promising potential. The ongoing development of suitable equipment, especially for continuous processing across various types of foods, and the standardization of process parameters are crucial for easy regulatory approval and improved food preservation.

This chapter aims to explore the science and technology involved in both traditional and innovative preservation methods.

Keywords: *Food Preservation, Microorganism, Shelf Life, Standardization, etc.*

Introduction

Food preservation is the use of scientifically based knowledge using a range of available technologies and techniques to prolong the shelf-life of food items and prevent deterioration and spoiling while guaranteeing that the product is free of harmful bacteria for customers. The amount of time it takes for a product to deteriorate to an undesirable degree is known as its shelf-life. Foods that have deteriorated will lose their flavour, texture, colour, and other sensory component. Nutritional quality is also affected during food deterioration. Food degradation may be caused by chemical, biological, microbiological, physical, or metabolic reasons. Prior to processing both plant- and animal-based foods, proper postharvest treatment should be implemented as early in the food production chain as feasible as part of preservation techniques. Processing processes often depend on suitable materials and ways of packing to ensure preservation throughout time. The way processed goods are handled throughout storage, transit, retail, and by the customer also affects how long they last. (Sancho, 2003).

1. Prevention or delay of microbial decomposition of food

- By keeping out micro-organisms (asepsis)
- By removal of micro-organisms (filtration)
- By hindering the growth or activity of micro-organisms (use of low temperature, drying, creating anaerobic conditions or using chemicals).
- By killing the micro-organisms (using heat or irradiation).

2. Prevention or delay of self -decomposition of food

- By destruction or inactivation of food enzymes (blanching or boiling)
- By prevention or delay of purely chemical reactions (use of antioxidants to prevent oxidation).

3. Prevention of damage by insects, animals, mechanical causes etc (use of fumigants, cushioning, packaging etc).

1. Prevention or delay of microbial decomposition

i) By keeping out micro-organisms (Asepsis): The goal of the asepsis procedure is to stop bacteria from getting into food by using either artificial or natural covers. Natural food barriers include things like eggshells, the skin or peel of fruits and vegetables (bananas, mangoes, citrus fruits, ash gourds, etc.), the outer shells of nuts (almonds, walnuts, and pecans), and the skin or fat in meat. Furthermore, packaging acts as a barrier to stop bacteria from getting into food. For example, keeping germs out of clean vessels in hygienic conditions and canning peaches or mushrooms helps avoid milk deterioration during collection and processing.

ii) By removal of micro-organisms (Filtration): Using bacteria-proof filters to filter liquid meals is a popular technique to remove all microorganisms from food. This method involves passing liquid meals through filters constructed of appropriate materials, such as asbestos pad, diatomaceous earth, unglazed porcelain, etc., and allowing them to percolate through with or without nano-filtration. Although they can be employed, centrifugation, sedimentation, cutting, washing, and other processes are not very efficient.

iii) By hindering the growth and activity of micro-organisms

a. **By using low temperature:** Foods kept at low temperatures are not as active in terms of microbiological development and enzyme activity. The food commodities can be kept at freezing temperatures (-18°C to -40°C) for frozen peas, frozen mushrooms, and root crops, or at chilling temperatures (0-50°C) for most fruits and vegetables, meat, poultry, fresh milk, and milk products.

b. **By drying of food commodity:** An essential part of food preservation is removing enough water so that microorganisms cannot thrive there. Moisture can be eliminated by using heat, as in mechanical and solar drying, or by binding the moisture and keeping it inaccessible to microorganisms by adding sugar (as in jams and jellies) or salt (high salt in raw mangoes). Osmotic dehydration, dried grapes (raisins), apricots, onions, cauliflower, etc. are a few examples.

c. **By creating anaerobic conditions:** Anaerobic environment can be produced by removing or venting the package's air or oxygen and replacing it with carbon dioxide or an inert gas such as nitrogen.

- Under such circumstances, no surviving bacteria or their spores can proliferate due to a lack of oxygen.
- The fermentation process produces carbon dioxide, which accumulates at the surface and anaerobically prevents the formation of aerobes.
- The same is achieved by carbonating beverages and refrigerating fresh food in controlled environments.
- This idea is demonstrated by canned food, where the food is sealed after the air is removed (exhausting).
- However, in order to keep the food from spoiling, all anaerobic bacteria and their spores must be eliminated.
- By keeping food from coming into contact with air, an oil coating on top of food inhibits the growth of microorganisms like yeasts and moulds.

d. **By use of chemicals:** When given in the right amounts, several chemicals can prevent food from spoiling by:

- Interfering with the microorganisms' cell membranes, their enzyme function, or their genetic mechanism.
- Through serving as an antioxidant.
- Higher concentrations of preservative might be hazardous to health; thus it is important to use the recommended amount as per the authorised rule.
- Chemical preservatives are benzoic acid and its sodium salt, sorbic acid, potassium meta-bi-sulphite, calcium propionates etc.
- Common antioxidants to check off flavour (rancidity) in edible oils include butyl hydroxy anisole (BHA), butyl hydroxy toluene (BHT), tertiary butyl hydroxy quinone (TBHQ), lecithin etc.
- Addition of organic acids like citric, acetic and lactic acid in the food inhibits the growth of many organisms.

iv) By killing the micro-organisms

a) **Use of heat:** Coagulation of proteins and inactivation of their metabolic enzymes by application of heat leads to destruction of micro-organisms present in foods. Foods that have been exposed to high temperatures also lose their enzyme content. Foods can be cooked to three different temperatures: 100°C for boiling, 100°C for pasteurisation, or 100°C for sterilisation.

i) **Pasteurization (heating below 100°C):** In food applications where extreme heat treatment results in unfavourable alterations, a gentle heat treatment is applied to eradicate the majority of harmful microorganisms. In order to increase shelf life, it is typically combined with other techniques. Generally speaking, pasteurisation is used as a low temperature long time (LTLT) or high temperature short time (HTST) method to milk and other dairy products.

- The LTLT procedure is the process of heating milk to 62.2°C for 30 minutes.
- The HTST method is defined as heating at 72 degrees Celsius for 15 seconds.
- Beer and grape wine are pasteurised at 60°C and 82–85°C, respectively, for one minute.
- The process of pasteurising juices, whether they are packed in bulk, bottles, or cans, relies on their acidity.
- Grape juice in bottles is pasteurised at 76.7°C for 30 minutes, while the juice in bulk is flash-heated to 80–85°C for a brief period of time.
- Bulk vinegar is kept at 60–65 °C for 30 minutes, while carbonated juice is cooked in bottles for 30 minutes at 65.6 °C.

ii) **Boiling (heating at 100°C):** Cooking of food including vegetables, meat etc by boiling with water involves a temperature around 100°C. Boiling of food at 100°C kills all the vegetative cells and spores of yeast and moulds and vegetative cells of bacteria.

- Many foods can be preserved by boiling (e.g. milk).
- Canning of acid fruit and vegetables (tomatoes, pineapple, peaches, cherries etc) is carried by boiling at about 100°C.
- Various terms used for heating of food are baking (in bread), simmering (incipient or gentle boiling), roasting (in meat), frying (shallow or deep fat frying) and warming up (small increase in temperature up to 100 °C).

iii) **Heating above 100°C:** A steam steriliser or retort is used to heat water under pressure to a temperature higher than 100 degrees Celsius. As steam pressure rises, the temperature within the retort also rises. The temperature in the retort increases with increase in steam pressure. The temperature in retort at mean sea level is 100°C; with 5psi pressure at 109°C; with 10psi pressure at 115.5°C and with 1 kg/cm^2 (100 Pa) pressure at 121.5°C.

- Mushrooms and other non-acid vegetables are canned at a temperature of 121.1°C under 15 psi of pressure.
- For sterilization of milk and other liquid foods like juices, ultra-high temperature (UHT) process is used.
- In UHT process, the food is heated to very high temperature (150°C) for only few seconds by use of steam injection or steam infusion followed by flash evaporation of the condensed steam and rapid cooling. The process is also used for bulk processing of many foods.

b) **Use of radiation:** To eliminate the microorganisms found in food, it is necessary to expose it to electromagnetic or ionising radiation. The use of UV lights to sterilise slicing blades in bakeries is one example of irradiation. Spices, fish, and a variety of other fruits have been exposed to gamma radiation from cobalt-60 or cesium-137 sources. They are also used to prevent potatoes and onions from sprouting.

2. Prevention or delay of self-decomposition of food

i) **By destruction or inactivation of food enzymes (blanching or boiling):** Before canning, freezing, or drying vegetables, blanching is a gentle heat treatment used to kill enzymes and stop food from decomposing naturally. Blanching is done by submerging the food item in boiling water or by leaving it in the steam for a short while, then quickly chilling it down.

ii) **By prevention or delay of purely chemical reactions (use of antioxidants to prevent oxidation):** Oxidation causes oils and fat-containing foods to get rancid and become unsafe to eat. Antioxidants such as lecithin, tertiary butyl hydroxy quinone (TBHQ), butyl hydroxy anisole (BHA), butyl hydroxyl toluene (BHT), and others are added in the right amounts to prevent oxidation and preserve food.

3. Prevention of damage by insects, animals, rodents and mechanical causes: Fumigants are used in cereals, dried fruits, and other food items to prevent insect and rat damage. Fruits are wrapped, padded trays are provided, light packs are used, and high-quality packing materials are used to prevent damage to fresh food items during handling and shipping.

Table 1: Methods of food preservation on the basis of food preservation principles.

Physical method	**Method**
By removal of heat (Preservation by low temperature)	Refrigeration, Freezing preservation, dehydro-freezing, carbonation
By addition of heat (preservation by high temperature	Pasteurization (LTLT, HTST), sterilization, UHT Processing, microwave.
By removal of water	Drying (open sun, solar/poly tunnel solar), Dehydration (mechanical drying), Evaporation/ concentration, Freeze concentration, reverse osmosis, freeze drying, foam mat drying and puff drying
By Irradiation	UV rays and gamma radiations
By non-thermal methods	High pressure processing, pulsed electric fields
Chemical methods	
a. By addition of acid (acetic or lactic)	Pickling (vegetable, olive, cucumber, fish, meat)
b. By addition of salt/brine	Salted mango/vegetable slices, salted and cured fish and meat i. Dry salting ii. Brining
c. By addition of sugar along with heating	Confectionary products like jams, jellies, preserves, candies, marmalades *etc*.
d. By addition of chemical preservatives.	i) Use of class II preservatives like Potassium meta-bi- sulphite, sodium benzoate, sorbic acid in food products. ii) Use of permitted and harmless substances of microbial origin like tyrosine, resin, niacin as in dairy products.
iii. By fermentation	i. Alcoholic fermentation (wine, beer) ii. Acetic acid fermentation (vinegar) iii. Lactic acid fermentation (curd, cheese, pickling of vegetables).
iv. By combination method	i. Combination of one or more methods for synergistic preservation. ii. Pasteurization combined with low temperature preservation. iii. Canning: heating combined with packing in sealed container. iv. Hurdle technology like low pH, salting, addition of acid, use of sugar, humectant and heating.

Source: Major food preservation techniques (Gould, 1989; Gould, 1995).

Methods of Food Preservation on the Basis of Food Preservation Principles

- **Addition of heat (or Thermal processing):** Thermal processing, sometimes referred to as applying heat, can be used to preserve food by deactivating enzymes and halting the development of microorganisms. Food may be preserved for a longer amount of time if it is packaged properly to prevent contamination. Pasteurisation involves the application of mild heat and is sometimes paired with freezing to produce a temporary extension of shelf life. On the other hand, shelf-stable products are produced by canning and other commercial sterilisation techniques. The use of heat treatment enhances the food's taste and safety characteristics during preparation.

- **Removal of heat (or cooling or refrigeration):** Heat removal, commonly achieved by refrigeration or chilling, is a commonly employed method for food preservation due to the temperature dependence of several physiological, biochemical, biological, and microbiological processes. Refrigeration is the most often used technique for maintaining food freshness. But while rotting processes persist, shelf life is only temporarily prolonged. On the other hand, freezing halts the majority of physiological and microbiological activities, save from a small number of chemical and enzymatic modifications. The product can be stored for an extended period of time using the freezing process provided that it is kept at or below -18°C.

- **Removal of moisture (or drying or dehydration):** Removal of moisture, achieved through methods like drying or dehydration, involves diminishing or eliminating the free moisture essential for life-sustaining activities in food. Food stability is accomplished by lowering the moisture content since most spoiling processes are either stopped or slowed down. This idea is used in many different processing techniques, including evaporation, concentration, and drying.

- **Regulating water activity:** Is important to keep food stable since food stability is influenced by moisture accessibility rather than just its presence. At a level of 0.75, water activity, a measure of accessible moisture, is considered adequate for the majority of activities. Salts, sugars, and other bigger molecules can bind moisture, making it inaccessible. Dried goods, foods with an intermediate moisture content, concentrates, and similar substances frequently have this characteristic.

- **Incorporating preservatives:** Acids, salt, and sugar all have different uses in different products. Preservatives have a specific function in controlling the actions of enzymes and microbes. Particularly, salt and sugar help control water activity. Some acids have antibacterial qualities, such the acetic acid in vinegar. These ideas are used to products like as jams, jellies, pickles, preserves, bottled drinks, and related goods.
- **Preservation by using chemicals:** Any material that has the ability to prevent, slow down, or stop the development of germs is considered a preservative. Chemical preservatives are another way to prevent microbiological deterioration of food products. The permeability of cell membranes, the activity of enzymes, and the process of cell division are all affected by preservatives, which have an inhibitory effect. Pasteurized squashes, cordials, and crushes have a cooked flavour. Especially in a tropical environment, they quickly ferment and deteriorate when the container is opened. Chemical preservatives are required to prevent this. Squashes and crushes that have been chemically preserved can be stored for a considerable time even after the bottle's seal has broken. However, because improper use of chemicals may be dangerous, it is imperative that their usage be adequately regulated. It is imperative that the preservative utilised is non-irritating and should not provide a health risk. It should be simple to identify and calculate. The FPO (1955) allows drinks to have two significant chemical preservatives (Sulphur dioxide and Benzoic acid).
    - **Sulphur dioxide**: It is extensively used all over the world to preserve a variety of products, including cordial, squash, crush, juice, pulp, and nectar. It inhibits enzymes, moulds, bacteria, and other preservation agents well. It also has bleaching and antioxidant properties. These characteristics aid in the retention of other oxidizable substances such as carotene and ascorbic acid. Additionally, it delays the onset of nonenzymatic discoloration or browning of the product. It is often utilised as its salts, including metabisulphite, bisulphate, and sulphite.
    - **Potassium metabisulphite ($K_2O_2SO_2$ (or) $K_2S_2O_5$)**: is frequently utilized to provide steady sources of SO_2. It is simpler to utilize than liquid or gaseous SO_2 since it is solid. It is fairly stable in neutral (or) alkaline media but decomposed by weak acids like carbonic, citric, tartaric acid, and malic acids. Additively, it combines with the acid in fruit or squash juice to generate potassium salt and release SO_2, which then mixes with the juice's water to form sulphurous acid.

Table 2: List of some common chemical preservatives

Chemical	Amount GRAS	Organism(s) affected	Use in Foods
Sulfites	200 - 300 ppm	Insects & microorganisms	Dried fruits, wine, lemon juice
Dehydroacetic acid	65 ppm	Insect	Strawberries
Sodium nitrite	120 ppm	Clostridia	Cured Meats
Ethyl formate	15 - 220 ppm	Yeast and molds	Dried fruits and nuts
Propionic acid	0.32%	Molds	Bread, cakes, cheeses
Sorbic acid	0.2%	Molds	Hard cheeses, cakes, salad dressings
Benzoic acid	0.1%	Yeast and molds	Margarine, relishes, soft drinks, ketchup, salad dressings

• Preservation of food by using heat:

Preserving food through high temperatures relies on the detrimental impact they have on microorganisms. High temperatures, in this context, encompass any temperatures above the ambient level. There are two commonly employed temperature categories for food preservation: pasteurization and sterilization.

- **Pasteurization,** achieved through heat application, involves either eliminating all disease-causing organisms (e.g., pasteurization of milk) or reducing the number of spoilage organisms in specific foods, such as the pasteurization of vinegar. The pasteurization of milk involves various temperature-time combinations, including 145°F (63°C) for 30 minutes (low temperature, long time [LTLT]), 161°F (72°C) for 15 seconds (primary high temperature, short time [HTST] method), 191°F (89°C) for 1.0 second, 194°F (90°C) for 0.5 second, 201°F (94°C) for 0.1 second, and 212°F (100°C) for 0.01 second. These treatments effectively eliminate heat-resistant non-sporeforming pathogenic organisms like Mycobacterium tuberculosis and Coxiella burnetii. Milk pasteurization temperatures also destroy yeasts, molds, gram-negative bacteria, and many gram-positives. The surviving organisms, after pasteurization, fall into two groups: thermodurics and thermophiles. Thermoduric organisms endure high temperatures but do not necessarily grow at these temperatures, often belonging to genera like Streptococcus and Lactobacillus. Thermophilic organisms, found in genera Bacillus and Clostridium, thrive and conduct metabolic activities at high temperatures. In brewing, pasteurization of beers, aimed at destroying spoilage biota, typically lasts for 8-15 minutes at 60°C.

- **Sterilization** entails the complete destruction of all viable organisms, as assessed through appropriate plating or enumerating techniques. Canned foods may be termed "commercially sterile" when cultural methods detect no viable organisms or when the survivors' numbers are so negligible as to be inconsequential under the conditions of canning and storage. Microorganisms present in canned foods may not grow due to unfavorable pH, oxidation-reduction potential (Eh), or storage temperature.

- **Preservation of food by using low temperature**

Preserving food through low temperatures relies on the principle that the activities of microorganisms associated with food can be decelerated above freezing temperatures and generally halted at subfreezing temperatures. This phenomenon is rooted in the fact that all metabolic reactions in microorganisms are enzyme-catalysed, and the rate of these reactions is temperature-dependent. As the temperature rises, there is a corresponding increase in the reaction rate, characterized by the temperature coefficient (Q_{10}). The Q_{10} can be defined as follows:

For most biological systems, the Q10 falls within the range of 1.5-2.5. Consequently, for every 10°C temperature increase within the suitable range, there is a twofold rise in the reaction rate. Conversely, with a 10°C decrease in temperature, the opposite holds true.

The term "psychrophile" was introduced by Schmidt-Nielsen in 1902 for microorganisms growing at 0°C. Over time, this term has been applied to organisms thriving in the subzero to 20°C range, with an optimal growth range of 10-15°C. Around 1960, the term "psychrotroph" was proposed for organisms capable of growing at 5°C or below. It is now widely accepted among food microbiologists that a psychrotroph is an organism growing at temperatures between 0°C and 7°C, forming visible colonies (or turbidity) within 7-10 days. Although some psychrotrophs can grow at temperatures as high as 43°C, they are, in fact, mesophiles. By these definitions, psychrophiles would be anticipated on products from oceanic waters or extremely cold climates. The microorganisms causing spoilage in the 0-5°C range are expected to be psychrotrophs.

There are Various Methods of Freezing

1. Sharp Freezing (Slow freezing): This technique, first used in 1861, involves freezing by circulation of air, either naturally or with the aid of fans. The temperature may vary from –15 to –29 °C and freezing may take from 3 to

72 hours. The ice crystals formed one large and rupture the cells. The thawed tissue cannot regain its original water content. The first products to be sharp frozen were meat and butter. Now-a-days freezer rooms are maintained at –23 to –29 °C or even lower, in contrast to the earlier temperature of –18° C

2. Quick freezing: In this process the food attains the temperature of maximum ice crystal formation (0 to – 4 °C) in 30 min or less. Such a speed results in formation of very small ice crystals and hence minimum disturbance of cell structure. Most foods are quick frozen by one of the following three methods

a) **By direct immersion**: Since liquids are good heat conductor's food can be frozen rapidly by direct immersion in a liquid such as brine or sugar solution at low temperature. Berries in sugar solution packed fruit juices and concentrates are frozen in this manner. The refrigeration medium must be edible and capable of remaining unfrozen at –18° C and slightly below. Direct immersion equipment's such as ottenson Brine freezer, Zarotschenzeff 'Fog' freezer, T.V.A. freezer, Bartlett freezer etc. of commercial importance earlier are not used today.

b) **By indirect contact:** The process of freezing a product by coming into contact with a metal surface that has been chilled by freezing brine or another refrigerating medium is known as indirect freeing. This is an outdated freezing technique where the food or package is kept in contact with the refrigerant flow at temperatures between –18 °C and –46 o C. This idea underpins the Knowles Automatic Package Freezer, Patterson Continuous Plate Freezer, FMC Continuous Can Freezer, and Birds Eye Freezers.

c) **By air blast**: In this method, refrigerated air at –18 °C to –34 °C is blown across the material to be frozen. The advantages claimed for quick freezing over slow freezing (sharp freezing) are:

(1) smaller (size) ice crystals are formed, hence there is less mechanical destruction of intact cells of the food.

(2) period for ice formation is shorter, therefore, there is less time for diffusion of soluble material and for separation of ice.

(3) more rapid preservation of microbial growth and

(4) more rapid slowing down of enzyme action.

3. **Cryogenic freezing:** While rapid freezing using the preceding methods preserves the quality of most products, ultrafast freezing is necessary for certain items. These materials undergo cryogenic freezing, which is the process

of freezing at extremely low temperatures (below –60° C). Liquid nitrogen and liquid CO_2 are now the refrigerants used in cryogenic freeing. In the former case, freezing may be achieved by immersion in the liquid, spraying of liquid or circulation of its vapour over the product to be frozen

4. Dehydro-freezing: This is a process where freezing is proceded by partial dehydration. In case of some fruits and vegetables about 50% of the moisture is removed by dehydration prior to freezing. This has been found to improve the quality of the food. Dehydration does not cause deterioration and dehydro frozen foods are relatively more stable. In this process food is first frozen at –18° C on trays in the lower chamber of a freeze drier and the frozen material dried (initially at 30°C for 24 hrs and then at 20°C). Under high vacuum (0.1 mm Hg) in the upper chamber. Direct sublimation of the ice takes place without passing through the intermediate liquid stage. The product is highly hygroscopic, excellent in taste and flavour and can be reconstituted readily. Mango pulp, orange juice concentrate, passion fruit juice and guava pulp are dehydrated by this method.

Traditional Methods for Food Preservation

The following are the main traditional methods for preservation of foods:

1. **Curing:** Reducing the moisture content in products like meat, fish, and vegetables through the osmosis process is the fundamental idea behind curing them. Reduced moisture level in food significantly reduces the possibility of microbial infection and subsequent growth. Curing is also used to add taste. To dehydrate the food, mixtures of salt, nitrates, sugar, and nitrites are added. Increased salinity during curing causes bacteria to become dehydrated, which kills them. Furthermore, salt can prevent rancidity by delaying the oxidation process, which in turn causes the fat to oxidise more slowly. (https://en.wikipedia.org/wiki/ Food preservation).

2. **Pickling:** There are other traditional methods, such as pickling—anaerobic fermentation that extends the shelf life of food—or submerging food in vinegar or vegetable oil. This process modifies the food's flavour, texture, and overall taste; the preserved goods are commonly referred to as pickles. Many Asian countries, including India, use pickles made from various vegetables, such as carrots, cauliflower, lemons, and raw mangoes. Pickles made from eggs, fish, and meat are consumed in many European nations as well as in Canada and the United States. Asian nations also often utilise the process of anaerobic fermentation, which involves keeping fruits and vegetables including carrots, radish, and mangoes in vinegar. Organic acids that serve as preservatives,

such as lactic acid and acetic acid, are produced during fermentation. Brine, or very salted water, is also employed as a preservation method in several nations. Bacteria and other microorganisms are killed by the circumstances. (Nummer, 2002)

3. **Canning:** French confectioner Nicolas Appert developed the use of canning to improve the shelf life of food items in the early 1800s. Cooking the food, sealing it in cans or jars, and boiling the sterilising containers are the steps in the procedure. Any remaining microbes are either killed off or become weaker under these conditions. Before Louis Pasteur demonstrated in 1864 the relationship between food spoilage and microbes, and hence illness, the procedure could not become widely accepted. (Nummer, 2002)
4. **Boiling:** Boiling water to destroy any microorganisms is a common method, particularly in impoverished nations. After that, it is cooled to normal temperature before being consumed. It's also common practice to boil milk—even pasteurised milk—before consuming it to eradicate any potential bacteria.
5. **Fermentation**: Specific bacteria are used in the fermentation process to produce several foods, including wine, cheese, and beer. By generating an acid or alcohol that is poisonous to other pathogenic microorganisms, these fermentative bacteria defend the meal against other pathogenic germs. Controlled elements like salt, temperature, oxygen content, and other factors are maintained throughout fermentation to aid the fermentative microbe in producing a food product fit for human consumption. (Nummer, 2002)

Modern methods for food preservation

The conventional method of heating food to a temperature that can destroy microorganisms is still in use today. After pasteurisation, milk is often sold on the market in sealed packages or jars. It is common in homes, particularly in Asian nations like India, to boil milk to destroy any potential microbes before storing it in a refrigerator to undergo pasteurisation.

1. **Freeze drying**: This is a contemporary method that uses suction to extract moisture contents from food at a considerably lower temperature while it's frozen. The fundamental idea is to sublimate solid water (ice) at lower pressure to evaporate it. It produces food that is of excellent quality. The shape of the foodstuff remains unchanged under the circumstances. Coffee is preserved and food is processed using this

method. Vacuum drying is also used for long-term storage of bacteria and yeasts. (Ratti, 2001)

2. **Vaccum packaging**: Using a vacuum pump to produce a vacuum, food is placed inside a plastic film bag and sealed once the bag has been vacuum-sucked. Microbes cannot develop in these conditions since they need oxygen to survive. This method preserves nuts' freshness without oxidising them, which is why packaging nuts is its primary application. (perdue, 2009).

Conclusion

The popularity of foods preserved using natural ingredients has grown because to consumer awareness of the harmful effects of synthetic chemical additions and growing concern about them. As a result, researchers—including food processors—are starting to choose organic preservatives more. Food additives must be natural, have minimal side effects and be widely accepted as safe if they are necessary for health reasons. In addition to keeping bacteria from proliferating, natural preservatives also extend the shelf life of many goods, including medicines and cosmetics. Additionally, it enables them to maintain their uniformity or freshness for extended periods of time without becoming hazardous. Artificial preservatives made of chemicals could be harmful to your health. The negative impacts of these pollutants, cosmetics, and medications are becoming more widely known. Compared to their synthetic counterparts, organic preservatives have a number of advantages, including the fact that they are also non-toxic and provide a number of health benefits.

References

Gould, G.W. 1989. Introduction Mechanisms of Action of Food Preservation Procedures.

Gould, G.W. 2000. Emerging technologies in food preservation and processing in the last 40 years. Innovations in food processing, 1-11.

Nummer, B.A. and Brian, A. 2002. Historical origins of food preservation. National Center for Home Food Preservation. University of Illinois Extension.

Ratti, C. 2001. Hot air and freeze-drying of high-value foods: a review. Journal of food engineering, 49(4): 311-319.

Trugo, L.C. and Finglas, P.M. 2003. Encyclopedia of food sciences and nutrition. L. C. Trugo and PM Finglas, Eds, 1498-1506.

Yam, K.L. ed. 2010. The Wiley encyclopedia of packaging technology. John Wiley & Sons.

8

Economics of Precision Livestock Farming

Saroj Bhati[1], Shriprakash Patel[2] and Ankit Kumar Maurya[3]

[1]Department of Animal Husbandry, CV&AS, SVPUAT, Meerut, Uttar Pradesh
[2]Department of Animal Husbandry and Dairy, CoA, CSAUAT, Kanpur Uttar Pradesh
[3]Department of Agricultural Economics and Statistics, CoA, CSAUAT Kanpur, Uttar Pradesh

Abstract

In the context of a rapidly growing global population, the livestock sector is grappling with increased demand for animal products and challenges like a shortage of fodder/feed resources and the consequences of climate change. To address these issues and improve production efficiency while mitigating environmental impact, various interventions are being explored. Among these, Precision Livestock Farming (PLF) tools are gaining prominence as a potential solution to bridge the demand-supply gap in the livestock industry. The evolution of PLF tools, primarily utilizing Information and Communication Technology (ICT), holds promise for enhancing productivity while prioritizing animal and human welfare. Through real-time monitoring of animal health, behavior, and environmental conditions, PLF facilitates timely decision-making and interventions. The transformative potential of PLF in revolutionizing the livestock sector is significant. However, successful implementation hinges on addressing adoption challenges, including technological barriers and ensuring widespread accessibility. Overcoming these hurdles not only facilitates the effective integration of PLF but also contributes to the sustainable development of the livestock industry amid global demands and environmental concerns.

Keywords: *Precision Livestock Farming (PLF), Automated Systems, Precision Feeding, Animal Health, etc.*

Introduction

Precision Livestock Farming (PLF) is a digitalized management system that continuously measures the production, reproduction, health and welfare of animals and environmental impacts of the herd by using information and Communication Technologies (ICT) and controls all stages of the production process. In conventional livestock management, decisions are mostly based on the appraisal, judgment and experience of the farmer, veterinarian and workers. The increasing demand for production and the number of animals makes it difficult for humans to keep track of animals. It is clear that a person is not able to continuously watch the animals 24 hours a day to receive reliable audio-visual data for management. Recent technologies already changed the information flow from animal to human, which helps people to collect reliable information and transform it into an operational decision-making process (e.g., reproduction management or calving surveillance). Today, livestock farming must combine requirements for a transparent food supply chain, animal welfare, health, and ethics as a traceable-sustainable model by obtaining and processing reliable data using novel technologies (Tekin *et al.*, 2021)

Additionally, the incorporation of innovative technologies in PLF not only addresses the challenges of human limitations in continuous monitoring but also fosters a shift towards proactive management. The utilization of sensor-based data collection and artificial intelligence algorithms enables early detection of anomalies in animal behavior and health, allowing for timely interventions and improved disease prevention strategies. This not only enhances animal welfare but also contributes to the overall efficiency of livestock operations. Furthermore, the advent of PLF aligns with the growing consumer demand for transparency in the food supply chain, as it facilitates the traceability of products back to their source, providing consumers with insights into the ethical and sustainable practices adopted in livestock farming. The integration of these advancements positions PLF as a key driver in the evolution of modern, responsible, and technologically-driven agriculture.

Concept of PLF (Precision Livestock Farming)

1. **Definition and Origin of PLF -** PLF integrates precision agriculture principles using sensors and actuators in large-scale livestock farming. Coined by Daniel Berckmans, PLF emphasizes continuous, real-time monitoring for rapid identification and control of issues related to animal health and welfare.

2. **Unified Perception of PLF-** PLF is described as a method for fine-grained management in modern livestock farming, utilizing process engineering, animal science and information technology. It is a set of technologies for real-time monitoring and control of animal health, welfare, production, reproduction, and environmental impacts.

3. **Management Goals of PLF-** PLF aims to provide stakeholders with real-time information for decision-making to improve the management of large-scale livestock and poultry. The ultimate goal is economically, socially and environmentally sustainable farming.

4. **Key Characteristics of PLF-** PLF relies on real-time data collection and analysis, with a core focus on intelligent sensing and analysis of individual animal information and behavior. It is a series of fine management methods supported by information technology to enhance animal productivity and welfare.

5. **Relationship with Smart Livestock Farming (SLF) and Digital Livestock Farming (DLF) -** SLF leverages information and communication technology (ICT) to manage cyber-physical livestock farms, with a focus on knowledge-based concepts. DLF incorporates precision and smart farming concepts, providing insights, modeling approaches and actionable analytics and automation techniques. Both SLF and DLF are considered as natural developments based on PLF, reflecting the increasing integration of digital technologies in animal husbandry.

6. **Development Beyond PLF-** DLF, in particular, focuses on transcending PLF by integrating precise data into digital systems, moving beyond mere accuracy. The development and application of these models are expected to improve productivity, quality and welfare in animal husbandry, promoting sustainability.

Uses and Importance of PLF

1. **Real-time Monitoring-** PLF enables continuous, real-time monitoring of animal behavior, health, and production metrics, providing immediate insights into their well-being.

2. **Early Disease Detection-** By analyzing data from PLF systems, farmers can detect early signs of diseases or abnormalities, allowing for timely intervention and improved disease management.

3. **Optimized Feeding-** PLF helps in precision feeding, ensuring that animals receive the right amount of nutrition tailored to their individual needs, promoting health and efficiency.

4. **Environmental Impact Mitigation-** Monitoring environmental parameters with PLF assists in minimizing the ecological footprint of livestock farming, contributing to sustainable and responsible agriculture.

5. **Efficient Resource Utilization-** PLF aids in optimizing the use of resources such as water, feed and energy, leading to more efficient and cost-effective livestock production.
6. **Enhanced Reproductive Management-** PLF provides insights into reproductive cycles, helping farmers optimize breeding programs and improve overall reproductive efficiency in livestock.
7. **Reduced Labor Intensity-** Automation in PLF reduces the need for manual labor in routine tasks, allowing farmers to focus on strategic decision-making and more specialized care.
8. **Improved Animal Welfare-** Continuous monitoring of animal health and behavior through PLF promotes better welfare practices, ensuring animals are kept in optimal conditions.
9. **Data-Driven Decision Making-** PLF generates a wealth of data, empowering farmers to make informed decisions about their livestock operations, leading to better management practices.
10. **Traceability in the Food Supply Chain-** PLF contributes to a transparent food supply chain by providing traceability from farm to table, meeting consumer demands for accountability and quality assurance.
11. **Precision in Medication Usage-** Through PLF, farmers can administer medications more precisely, reducing the risk of antibiotic resistance and ensuring responsible use of pharmaceuticals.
12. **Early Calving Surveillance-** PLF systems can aid in monitoring pregnant animals, providing alerts for imminent calving, allowing for timely assistance and reducing calving-related issues.

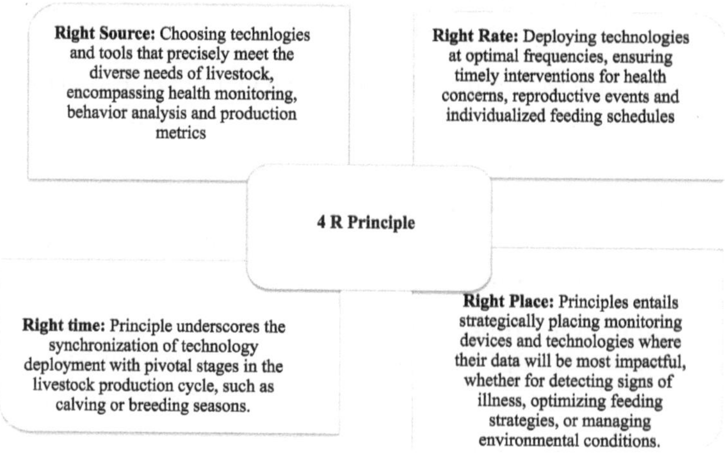

Fig.4'R Principle of Precision Farming

Precision Technologies in Livestock Farming

Pedometer- The utilization of an automated system known as a pedometer proves instrumental in monitoring walking activity and milk production in dairy farms. As highlighted by Edwards and Tozer (2004), this tool allows for the observation of various daily movements in dairy cows, including milking, eating, standing and lying, while also detecting changes in these activities. Presently, the primary application of daily walking activity monitoring is in the detection of estrus among dairy cows. Additionally, a reduction in daily walking activity, coupled with a decrease in milk yield, serves as an early warning signal for potential disorders in dairy cattle. Edwards and Tozer (2004) reported that by closely monitoring daily walking activity, disorders such as ketosis and digestive issues in fresh cows could be identified 7 to 8 days earlier, compared to a 5 to 6 days earlier detection based on milk yield alone. Consequently, daily walking activity emerges as a valuable tool for the early detection of transition cow disorders, contributing to the prevention of further reduction in milk yield. In a related study, (Mazrier *et al.*, 2006) demonstrated that pedometers were effective in predicting lameness earlier than the appearance of clinical signs in a dairy cow herd. This was achieved by correlating pedometric activity (PA) with clinical cases of lameness.

GPS- The adoption of GPS 'collars' for livestock, including dairy animals, has become prevalent over the past two decades. This technology enables the detailed recording of position information over extended periods, offering a comprehensive understanding of the habits and spatial distribution factors affecting ruminants (Laca, 2009). Such systems would provide insights into the real-time location and activities of all animals, enhancing overall management capabilities.

Biometric- Biometric based identification in cattle relies on the muzzle point image pattern, which serves as a primary characteristic for individual recognition, akin to a human fingerprint. In ruminants, the vascular pattern in the muzzle is distinct for each animal. Additionally, the scanning of the eye and recording of the retinal image contribute to the uniqueness of biometric identification in cattle.

Automatic milking systems - Automatic milking systems offer the unique feature of allowing cows to voluntarily visit the milking robot. This results in a considerable variation in milking intervals due to the voluntary nature of cow visits (Hogeveen *et al.,* 2001). The key components of an automatic milking system include the milking stall, teat cleaning system, teat detection system, robotic arm for teat cup attachment, control system with sensors and software and the milking machine itself.

The advantages of automatic milking systems include increased milk yield due to more frequent milking (De Koning and C.J.A.M., 2010). Various precision dairy farming technologies complement automatic milking, encompassing daily milk yield recording, monitoring milk components (such as fat, protein, and somatic cell count), pedometers, automatic temperature recording devices, milk conductivity indicators, automatic estrus detection monitors, and daily body weight measurements (Bewley, J.M., 2010).

Activity monitoring- Despite the widespread use of hormonal synchronization protocols enabling timed artificial insemination (TAI), the identification of behavioral estrus remains crucial in overall reproductive management (Caraviello *et al.,* 2006 and Miller *et al.,* 2007). Given the impact of AI service rate on reproductive performance and challenges with visual estrus detection, modern electronic systems integrate activity monitoring (Holman *et al.,* 2011 and Jonsson *et al.,* 2011). These systems associate heightened physical activity with estrous behavior in cattle, leveraging increased restlessness and general physical activity during estrus (Farris, 1954).

Rumination Collars- Rumination collars operate by identifying rumination through the analysis of chewing sounds, specifically excluding those associated with eating. The parameter of rumination time proves valuable in two key areas: nutrition management and the health and welfare sensing of cows. Scientific literature widely acknowledges that, on average, a dairy cow spends approximately 35-40% of the day engaged in rumination.

Reticulo rumen boluses- Reticulo-rumen boluses are equipped with specialized sensors (such as temperature, pH, and pressure) designed to evaluate rumen function. These sensors can be particularly relevant for detecting Sub Acute Ruminal Acidosis (SARA) by monitoring rumen pH and assessing feeding and drinking behavior (Caja *et al.,* 2016). The advancement of wireless telemetry in the rumen has enhanced the monitoring of metabolic disorders, including sub-acute ruminal acidosis. Additionally, automated weighing and the promising development of body condition scoring (BCS) through image capture contribute to more comprehensive metabolic disorder monitoring (Davies, 2016).

Precision Farming Technologies in Pig Farming

Precision technologies have become integral in modern pig farming, offering a multifaceted approach to enhance various aspects of management. These innovations include the assessment of pig welfare through sophisticated sound analysis and continuous monitoring of their behavior. Additionally, the use of automatic body weight estimation via advanced image analysis provides valuable insights into the health and growth of the pigs. The implementation of precision technologies goes beyond the livestock, extending to the management of waste generated within the farm, promoting more sustainable and environmentally conscious practices. Moreover, a comprehensive slaughterhouse registration system is integrated, ensuring traceability and accountability throughout the entire farming process. Together, these precision technologies contribute to elevated efficiency, informed decision-making, and improved overall outcomes in pig farming.

Real-time sound analysis- Real-time sound analysis plays a pivotal role in addressing respiratory pathologies prevalent in intensive pig farming. These conditions often manifest through frequent coughing, a prominent clinical sign (Islam *et al.,* 2013). Pig vocalization has been linked directly to pain, prompting attempts to classify these sounds (Marx *et al.,* 2003). Veterinarians commonly rely on assessing cough sounds during farm visits for diagnostic purposes. However, the limitations of manual assessment and the need for continuous monitoring have led to the development of automated tools (Moreaux *et al.,* 1999 & Van Hirtum and Berckmans 2002). Implementing an automatic monitoring system for animals' coughs enables real-time evaluation without the constraints of on-site visits. This innovation not only facilitates early detection of respiratory issues but also contributes to enhanced farm management by enabling timely and targeted treatments (Silva and Exadaktylos, 2009).

Body weight estimation by image analysis- In the context of body weight estimation in pig farming, a top-view camera was employed to monitor each pen. The system automatically identified individual pigs based on their distinct painting patterns using shape recognition techniques. The weight estimation process involved several steps. Initially, an ellipse fitting algorithm was applied to localize pigs in the image. Subsequently, the area occupied by the pig within the ellipse was calculated. Finally, the weight of the pigs was estimated using dynamic modeling. To validate the accuracy of the model, the estimated weights were compared against manually recorded actual weights obtained through biweekly measurements for each individual pig. This comprehensive approach ensures a reliable and efficient means of monitoring and estimating the body weight of pigs in a farming environment (Kashiha *et al.,* 2014)

Precision Feeding

Precision nutrition in livestock farming aims to precisely meet the nutrient needs of animals for safe, high-quality, and efficient production while minimizing the environmental impact. Unlike a population-based approach, nutrient requirements are treated as individual statistics for each animal (Banhazi *et al.*, 2014). Precision feeding is a crucial strategy to enhance nutrient utilization, reduce feeding costs, and minimize nutrient excretion (Pomar *et al.,* 2011). It holds great promise in mitigating greenhouse gas (GHG) and ammonia emissions (Gerber *et al.,* 2013).

The proposed sustainable precision livestock farming system incorporates real-time collection of individual feed intake and body weight information. This data is then utilized to estimate optimal nutrient concentrations in diets for each pig within the herd, employing innovative modeling approaches. This comprehensive strategy aligns with sustainable practices and enhances the overall efficiency of livestock farming (Hauschild *et al.,* 2012).

Activity Monitoring

1. RFID Tags

a. Purpose - Individual animal identification.

b. Uses - Tracking, monitoring, and managing individual animals.

c. Benefits - Improved traceability, efficient management.

2. Electronic Health Monitoring Systems -

a. Purpose - Continuous health monitoring.

b. Uses - Early disease detection, intervention.

c. Benefits - Timely treatment, improved welfare.

3. Automated Temperature Recording Devices -

a. Purpose - Monitoring environmental conditions.

b. Uses - Heat stress detection, comfort assessment.

c. Benefits - Health optimization, productivity.

4. Mastitis Detection Systems

a. Purpose - Early mastitis detection.

b. Uses - Monitoring udder health.

c. Benefits - Reduced antibiotic use, improved milk quality.

5. Automatic Estrus Detection Monitors -

a. Purpose - Estrus identification.

b. Uses - Optimal breeding timing.

c. Benefits- Increased reproductive efficiency.

6. Herd Navigator -

a. Purpose - Advanced milk analysis.

b. Uses - Heat detection, mastitis, ketosis monitoring.

c. Benefits - Comprehensive health insights..

7. Automated Water Quality Monitoring -

a. Purpose- Monitoring water conditions.

b. Uses - Ensuring water quality, hydration tracking.

c. Benefits - Health optimization, production.

8. Automated Culling Systems -

a. Purpose - Identifying animals for culling.

b. Uses - Optimal herd management.

c. Benefits - Improved genetics, resource efficiency.

9. Smart Feeders

a. Purpose- Controlled feeding systems.

b. Uses- Individualized nutrition, weight management.

c. Benefits - Resource optimization, health improvement.

10. Smart Barn Sensors-

a. Purpose - Monitoring barn conditions.

b. Uses - Temperature, humidity control.

c. Benefits - Comfort optimization, health.

11. Behavioral Analytics Software -

a. Purpose - Analyzing animal behavior.

b. Uses- Stress detection, disease patterns.

c. Benefits - Early intervention, improved welfare.

12. Automated Medication Dispensers

a. Purpose - Controlled medication administration.

b. Uses- Timely treatment, disease management.

c. Benefits- Reduced labor, precise dosing.

d. Constraints - Calibration, cost.

13. Smart Ear Tags

a. Purpose - Health and location monitoring.

b. Uses - Individual tracking, health alerts.

c. Benefits - Enhanced traceability, early intervention.

14. Automatic Calf Feeders -

a. Purpose - Automated calf feeding.

b. Uses - Calf health, growth monitoring.

c. Benefits - Reduced labor, precise nutrition.

15. Automated Fly Control Systems

a. Purpose - Fly prevention for livestock.

b. Uses - Animal welfare, disease prevention.

c. Benefits - Improved comfort, health.

It's important to note that while these tools offer numerous benefits, their implementation may come with challenges such as initial costs, technical complexities and the need for ongoing maintenance and calibration. Additionally, ethical considerations should be taken into account, especially concerning automated procedures and animal welfare.

Automated Feeding Systems

In Precision Livestock Farming (PLF), automated feeding systems play a crucial role in optimizing the feeding process for livestock. Here's a step-by-step overview of how automated feeding works in PLF, along with a suitable example-

Step 1: Data Collection and Monitoring- Automated feeding systems start with the collection of real-time data. Sensors and monitoring devices are employed to gather information on individual animals' characteristics, such as weight, health status, and feeding behavior. Example- Sensors in the feeding area track each animal's presence, feeding frequency, and amount consumed.

Step 2: Individualized Animal Profiles- The collected data is used to create individualized profiles for each animal within the herd. These profiles include factors like age, weight, production potential and specific nutritional requirements. Example- An automated system assigns a unique profile to each cow, considering factors like age, milk production stage, and health status.

Step 3: Nutrient Requirements Assessment- Precision feeding involves evaluating the nutritional potential of feed ingredients and determining the specific nutrient requirements for each animal based on its profile. Example- The system calculates the ideal balance of proteins, carbohydrates and other nutrients needed for optimal health and performance.

Step 4: Formulation of Customized Diets- Automated feeding systems use the assessed nutrient requirements to formulate customized diets for each animal. These diets are designed to meet the individual nutritional needs identified in their profiles. Example- A robotic feeding system combines different feed ingredients in precise proportions to create a customized mix for each cow.

Step 5: Real-time Adjustments- Continuous monitoring allows the system to make real-time adjustments to the feed composition and quantity based on changes in an animal's condition or external factors. Example- If a cow shows signs of illness or increased milk production, the system can automatically adjust the nutrient concentration in its feed.

Step 6: Feed Delivery- The automated feeding system delivers the customized feed to each animal according to its profile. This can involve conveyors, robotic arms, or other mechanisms to ensure accurate and timely distribution. Example- Robotic feeders dispense individualized portions directly to each animal, promoting fair access to feed.

Step 7: Performance Evaluation- Over time, the system evaluates the performance of each animal based on health indicators, productivity and overall well-being. Example- Data on milk production, weight gain, and health metrics are continuously analyzed to assess the effectiveness of the precision feeding strategy.

How does a PLF system work?

A Precision Livestock Farming (PLF) system typically involves the integration of various technologies and sensors to monitor and collect data related to the health, behavior and performance of individual animals or the entire herd. The primary goal is to provide farmers with accurate and real-time information, enabling informed decision-making and improved management practices. Here's a general overview of how a PLF system works and how farmers can leverage the data-

1. Data Collection-

Tools Used

- Sensors and Wearables- Devices like RFID tags, GPS collars and smart ear tags monitor individual animal movements, location and activity.
- Health Monitoring Systems- Sensors measure physiological parameters, detect changes in behavior and identify signs of distress or illness.
- Automated Milking Systems- Record milk yield, composition and analyze milking patterns.
- Environmental Sensors - Monitor barn conditions, including temperature, humidity and air quality.

2. Data Transmission and Storage- Collected data is transmitted to a central system, often cloud-based, where it is securely stored and accessible for analysis.

3. Data Analysis and Interpretation

Tools Used

- Data Analytics Software - Processes the collected data to identify patterns, trends and anomalies.
- Artificial Intelligence (AI) and Machine Learning- Enables predictive analytics and identifies correlations between various factors affecting livestock.

4. Alerts and Notifications- PLF systems can generate real-time alerts or notifications when anomalies or deviations from normal patterns are detected. This helps farmers promptly address potential health issues, optimize feeding regimes, or identify reproductive opportunities.

5. Presentation to the Farmer

Mobile Apps- Some PLF systems offer mobile applications, allowing farmers to access real-time data and receive alerts on their smartphones or tablets, enabling remote monitoring.

Dashboard Interfaces- The analyzed data is presented to farmers through user-friendly dashboard interfaces. These interfaces provide visual representations of key metrics, trends, and alerts.

Processed data is presented to the farmer through user-friendly dashboards or applications, offering insights into -

- Individual animal health status.
- Behavior patterns indicating stress, estrus, or illness.
- Production metrics, including milk yield and composition.
- Environmental conditions affecting livestock well-being.

6. Decision-Making - Armed with the insights provided by the PLF system, farmers can make informed decisions. For example

- Health Management- Administering targeted treatments based on early disease detection.
- Reproductive Optimization- Timely intervention for artificial insemination or breeding.
- Nutritional Adjustments - Modifying feeding strategies based on individual animal needs.
- The farmer interprets the presented data and makes informed decisions based on the insights gained.
- Decisions may include adjusting feeding regimes, administering veterinary care, optimizing breeding schedules, or implementing changes in environmental conditions.

7. Continuous Monitoring and Iteration- PLF systems enable continuous monitoring, creating a feedback loop for ongoing improvement. Farmers can adjust strategies, refine management practices and optimize resources based on the evolving data and insights.

8. Feedback Loop- Continuous monitoring and feedback loops ensure that the system adapts to changes and refines its predictions over time.

Farmers can manually input additional information or observations to enhance the system's understanding.

9. Integration with Other Systems- PLF systems can be integrated with other farm management systems, such as feeding systems, robotic milking machines, or environmental controls. This integration enhances overall farm efficiency and coordination.

By leveraging the data provided by PLF systems, farmers can enhance productivity, improve animal welfare and optimize resource utilization. The ability to make data-driven decisions contributes to more sustainable and efficient livestock farming practices.

Precision Livestock Farming

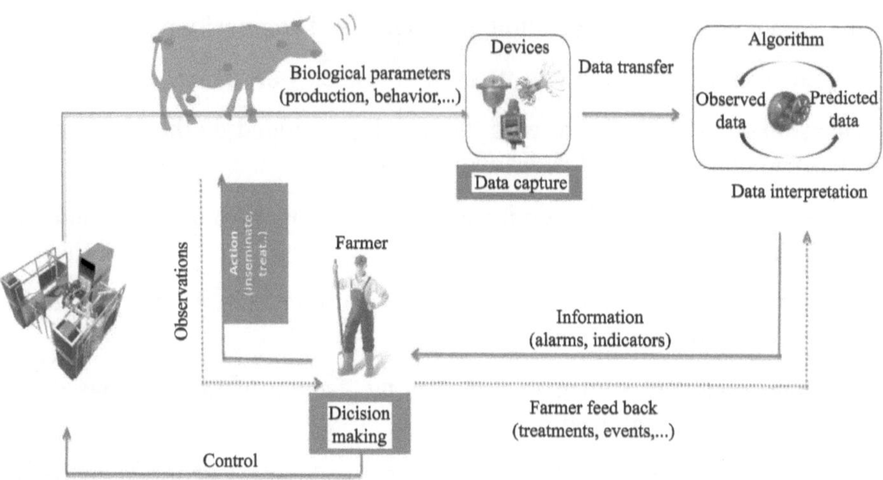

Source: www.google.com

Adopting Precision Livestock Farming (PLF) in India Faces Several Challenges and Problems

1. **High Initial Investment -** The cost of implementing PLF technologies, including sensors, automated systems and data management tools, can be prohibitive for many farmers in India.

2. **Limited Awareness and Education -** Many farmers lack awareness and education about the benefits and functionalities of PLF, hindering their willingness to adopt these technologies.

3. **Infrastructure Gaps -** Inadequate rural infrastructure, including unreliable electricity supply and poor internet connectivity, poses significant challenges for the deployment and operation of PLF systems.

4. **Diversity in Livestock Farming Practices -** India has diverse livestock farming practices and adapting PLF solutions to suit the varied needs of different regions and types of livestock can be challenging.

5. **Lack of Technical Support -** A shortage of technical expertise and support services for PLF technologies may leave farmers without the necessary guidance for successful implementation.

6. **Data Privacy Concerns-** Farmers may have concerns about the privacy and security of the data collected by PLF systems, especially in the absence of robust data protection regulations.

7. **Resistance to Change-** Traditional farming practices are deeply ingrained sand there may be resistance to adopting new technologies due to cultural, social, or psychological factors.
8. **Small Landholdings-** The prevalence of small and marginal landholdings in India may limit the scalability of PLF technologies, which are often designed for larger operations.
9. **Water Scarcity-** In regions facing water scarcity, the environmental monitoring aspects of PLF may be constrained, affecting accurate data collection and analysis.
10. **Market Access for PLF Data-** Farmers may struggle to find markets for the data generated by PLF systems, limiting opportunities for income generation through data monetization.
11. **Policy and Regulatory Challenges-** Lack of clear policies and regulations governing the use of PLF technologies may create uncertainty and inhibit farmer confidence in adopting these systems.
12. **Access to Credit-** Limited access to credit for farmers, especially smallholders, can impede their ability to invest in PLF technologies, which often require significant upfront investment.
13. **Technological Interoperability-** Lack of standardization and interoperability between different PLF technologies may create compatibility issues, limiting the seamless integration of systems.
14. **Climate Variability-** The unpredictable climate in many parts of India can affect the reliability of PLF data, especially in instances of extreme weather events.
15. **Limited Extension Services-** The absence of effective extension services and outreach programs may hinder the dissemination of information and knowledge about PLF among farmers.

Addressing these challenges requires a concerted effort from policymakers, technology developers and the agricultural community to create an enabling environment for the successful adoption of PLF in India.

Overcoming the challenges associated with the adoption of Precision Livestock Farming (PLF) in India requires a multifaceted approach involving various stakeholders. Here are strategies to address some of the key issues

1. **Financial Support and Subsidies-** Government Initiatives- Implement schemes and subsidies to provide financial support to farmers, especially smallholders, for the acquisition and implementation of PLF technologies.
2. **Education and Awareness Programs-** Training Programs- Conduct training sessions and workshops to educate farmers about the benefits and functionalities of PLF technologies, emphasizing how they can enhance productivity and profitability.
3. **Infrastructure Development-** Rural Infrastructure Improvement- Invest in improving rural infrastructure, including electricity supply and internet connectivity, to ensure the reliable operation of PLF systems.
4. **Tailored Solutions for Diverse Practices-** Research and Development- Invest in R&D to develop PLF solutions that are adaptable to diverse livestock farming practices in different regions of India.
5. **Extension Services-** Enhanced Extension Services- Strengthen extension services to provide continuous support and guidance to farmers in the adoption and utilization of PLF technologies.
6. **Data Security Measures-** Regulatory Framework- Develop and implement clear regulations and policies regarding data privacy and security to address concerns and build trust among farmers.
7. **Collaboration and Partnerships-** Public-Private Partnerships- Foster collaborations between government agencies, private enterprises and research institutions to jointly address challenges and promote the adoption of PLF.
8. **Water Management Strategies-** Water Conservation Technologies- Integrate water management technologies with PLF systems to address water scarcity issues and improve overall resource efficiency.
9. **Inclusive Policies-** Inclusive Policies- Develop policies that cater to the needs of small and marginal farmers, ensuring that they can also benefit from PLF technologies.

10. **Climate-Resilient Technologies-** Adaptation Strategies- Develop PLF technologies that are resilient to climate variability and extreme weather events, ensuring reliable operation in diverse environmental conditions.
11. **Market Access for Data-** Data Monetization Platforms- Facilitate the creation of platforms or marketplaces where farmers can monetize their PLF-generated data, creating additional income streams.
12. **Capacity Building-** Continuous Training- Establish ongoing training programs to continually build the capacity of farmers in PLF technology use and management.

Threats of Precision Livestock Farming for Animal Welfare

1. Technical Glitches within the Precision Livestock Farming (PLF) System

a) **System Downtime-** Technical failures, such as power cuts, software bugs, or hardware breakdowns, can lead to unplanned system downtime. This downtime may disrupt the continuous monitoring and data collection crucial for precision livestock farming.

b) **Data Inaccuracy-** Technical malfunctions may result in inaccurate data collection and transmission. This could lead to false alerts or failures in recognizing actual issues within the livestock, impacting the reliability of decision-making based on the PLF system.

c) **Environmental Vulnerability-** PLF hardware is often exposed to harsh farm environments, including high dust concentrations and the presence of non-target animals. These conditions can compromise the integrity of sensors and obstruct their functionality, contributing to technical failures.

2. Impact of the PLF System on Animals

a) **Behavioral Stress-** Introduction of PLF systems, such as automatic milking or feeding systems, may induce stress in animals during the learning phase or adaptation to new technologies. Behavioral changes, like fear or anxiety, could affect the overall well-being of the animals.

b) **Pathogen Transmission Risk-** Certain PLF technologies, like automated milking systems, have been associated with increased transmission of pathogens. Higher bulk tank somatic cell counts and the potential for increased disease prevalence can impact the health of animals within the system.

3. Insufficient External Validation of PLF System

a) **Unreliable Alerts-** Poor external validation may lead to unreliable alerts generated by PLF algorithms. This can result in false positives or false negatives, causing caretakers to make incorrect decisions based on inaccurate information about animal behavior or health.

b) **Limited Performance Under Real Conditions-** PLF systems might perform well under controlled testing conditions but may struggle to maintain accuracy in real-life, dynamic farm environments. Factors such as variability in animal behavior and environmental conditions could compromise the system's reliability.

c) **Algorithmic Bias-** Inadequate external validation may contribute to algorithmic bias, where PLF systems reproduce and reinforce existing biases present in the training data. This can impact the accuracy of alerts and potentially lead to inappropriate responses to animal welfare issues.

4. Reduced Interaction Time Between End Users and Animals in PLF Systems

a) **Negative Impact on Human-Animal Bond-** A reduction in the time spent with animals may negatively impact the human-animal bond. The quality of interactions, including observation, care, and attention to individual needs, may diminish, affecting the overall well-being and welfare of the animals.

b) **Focus on Efficiency Over Animal Welfare-** The quest for increased production efficiency may prioritize tasks that can be automated, such as feeding or monitoring, at the expense of more nuanced and personal interactions. This shift may lead to a more transactional approach to animal care.

c) **Decreased Knowledge of Individual Animals-** Less time spent per animal can result in a reduced understanding of individual animals, including their personalities and health conditions. Farmers may be less likely to detect subtle signs of distress or illness, impacting the early identification of welfare issues.

5. Adapting housing and management to Precision Livestock Farming (PLF) systems instead of focusing on the well-being of animals can lead to several concerns

a) **Inadequate Animal Welfare-** Designing facilities primarily for PLF technologies without considering the comfort and natural behavior of the animals may lead to poor living conditions, impacting their overall welfare.

b) **Behavioral Restrictions-** Housing and management practices tailored solely to accommodate PLF systems might restrict the natural behaviors of animals, such as grazing, social interactions, and movement, which are crucial for their health and well-being.

c) **Increased Stress Levels-** If housing structures are not optimized for animal comfort but rather for efficient PLF implementation, it can contribute to elevated stress levels among animals, negatively affecting their health and productivity.

6. The increased instrumentalization of animals due to the reciprocal strengthening effects between Precision Livestock Farming (PLF) and the intensification of animal production can have several implications

a) **Loss of Individual Identity-** The industrialization and intensification associated with PLF may lead to large-scale, confined farming practices, making it challenging for society to view farm animals as individuals with unique personalities and characteristics.

b) **Objectification of Animals-** Animals may be treated as commodities rather than sentient beings, fostering a mindset that prioritizes economic efficiency over the welfare and ethical treatment of individual animals.

7. The promotion of animal consumption and harm facilitated by Precision Livestock Farming (PLF) can lead to various consequences

a) **Increased Meat Consumption-** PLF systems, often designed to enhance production efficiency, may contribute to an increase in the availability and affordability of animal products. This could result in higher consumption levels globally.

b) **Diminished Concern for Animal Welfare-** Higher meat consumption is linked to reduced empathy and concern for animal welfare. As more people consume animal products, there may be a societal shift towards accepting practices that prioritize production efficiency over ethical considerations.

c) **Moral Dissonance-** The disconnect between the consumption of animal products and the awareness of animal exploitation may create cognitive dissonance. Individuals may downplay the moral status of animals to alleviate the ethical conflict associated with consuming products derived from sentient beings.

Conclusion

Precision Livestock Farming (PLF) stands at the forefront of transformative changes in the agricultural landscape, offering immense potential for improving the efficiency, sustainability and welfare of livestock farming. By leveraging cutting-edge technologies such as sensors, data analytics and automation, PLF provides farmers with unprecedented insights into individual animal health, behavior, and performance. The real-time monitoring and precision management offered by PLF contribute to early disease detection, optimized feeding strategies and enhanced reproductive efficiency. While the promises of PLF are substantial, its adoption is not without challenges. Issues such as high initial costs, limited awareness and diverse farming practices pose hurdles that require collaborative efforts from policymakers, researchers and the agricultural community to overcome. The integration of PLF into traditional farming practices necessitates targeted education, infrastructure development and supportive policies to ensure inclusivity and equitable access. As PLF evolves, it is essential to address ethical considerations related to data privacy, animal welfare and the potential depersonalization of the farmer-animal relationship. Additionally, the ongoing development of standardized protocols and interoperability among different PLF systems will foster seamless integration and enhance the overall impact of these technologies. In essence, Precision Livestock Farming holds the promise of ushering in a new era of smart and sustainable agriculture. By embracing innovation, addressing challenges and fostering a culture of continuous learning, the agricultural sector can unlock the full potential of PLF to meet the demands of a growing population, ensure food security and contribute to the long-term sustainability of livestock farming.

References

Banhazi, T. M., Lehr, H., Black, J. L., Crabtree, H., Schofield, P., Tscharke, M., & Berckmans, D., 2012. Precision livestock farming: an international review of scientific and commercial aspects. International Journal of Agricultural and Biological Engineering, 5(3): 1-9.

Bewley, J., 2010. Precision dairy farming: Advanced analysis solutions for future profitability. In The First North American Conference on Precision Dairy Management (Vol. 16).

Caja, G., Castro-Costa, A., & Knight, C. H., 2016. Engineering to support wellbeing of dairy animals. Journal of Dairy Research, 83(2): 136-147.

Davies, D., 2016. Precision dairy farming: a review of current available technologies, Knowledge Exchange Hub, IBERS, Aberystwyth University

De Koning, C. J. A. M., 2010. Automatic milking–common practice on dairy farms.

Edwards, J. L., & Tozer, P. R., 2004. Using activity and milk yield as predictors of fresh cow disorders. Journal of dairy science, 87(2): 524-531.

Hauschild, L., Lovatto, P. A., Pomar, J., & Pomar, C., 2012. Development of sustainable precision farming systems for swine: estimating real-time individual amino acid requirements in growing-finishing pigs. Journal of animal science, 90(7): 2255-2263.

Hogeveen, H., Ouweltjes, W. C. J. A. M., De Koning, C. J. A. M., & Stelwagen, K., 2001. Milking interval, milk production and milk flow-rate in an automatic milking system. Livestock production science, 72(1-2): 157-167.

Jónsson, R., Blanke, M., Poulsen, N. K., Caponetti, F., & Højsgaard, S., 2011. Oestrus detection in dairy cows from activity and lying data using on-line individual models. Computers and electronics in agriculture, 76(1): 6-15.

Kashiha, M., Bahr, C., Ott, S., Moons, C. P., Niewold, T. A., Ödberg, F. O., & Berckmans, D., 2014. Automatic weight estimation of individual pigs using image analysis. Computers and Electronics in Agriculture, 107: 38-44.

Laca, E. A., 2009. Precision livestock production: tools and concepts. Revista brasileira de zootecnia, 38: 123-132.

Marx, G., Horn, T., Thielebein, J., Knubel, B., & Von Borell, E., 2003. Analysis of pain-related vocalization in young pigs. Journal of sound and vibration, 266(3): 687-698.

Mazrier, H., Tal, S., Aizinbud, E., & Bargai, U., 2006. A field investigation of the use of the pedometer for the early detection of lameness in cattle. The Canadian Veterinary Journal, 47(9): 883.

Moreaux, Beerens, & Gustin., 1999. Development of a cough induction test in pigs: effects of SR 48968 and enalapril. Journal of Veterinary Pharmacology and Therapeutics, 22(6): 387-389.

Pomar, C., Hauschild, L., Zhang, G. H., Pomar, J., & Lovatto, P. A., 2009. Applying precision feeding techniques in growing-finishing pig operations. Revista Brasileira de Zootecnia, 38: 226-237.

Silva, M., Exadaktylos, V., Ferrari, S., Guarino, M., Aerts, J. M., & Berckmans, D., 2009. The influence of respiratory disease on the energy envelope dynamics of pig cough sounds. Computers and electronics in agriculture, 69(1): 80-85.

Van Hirtum, A., & Berckmans, D., 2002. Assessing the sound of cough towards vocality. Medical Engineering & Physics, 24(7-8): 535-540.

9

Profitability of Crop Rotation System

Mandeep Kumar[1], Naushad Khan[2], Sandeep Kumar Maurya[3], Shravan Kumar Maurya[4] and Durgesh Kumar Maurya[5]

[1,2,4&5]*Department of Agronomy, CoA, CSAUAT, Kanpur, Uttar Pradesh*

[3]*Department of Agronomy, Rama University, Kanpur, Uttar Pradesh*

Abstract

Crop rotation is a sustainable agricultural practice that involves systematically planting different crops in a specific sequence on the same piece of land. This abstract explores the profitability of implementing a crop rotation system. The benefits of crop rotation include improved soil fertility, pest and disease control, and enhanced nutrient cycling. These factors contribute to increased crop yields and overall farm productivity. One key aspect of profitability in a crop rotation system is the reduction of input costs. By diversifying crops, farmers can minimize the need for chemical fertilizers and pesticides, leading to lower production expenses. Additionally, the practice helps mitigate the risk of crop failure associated with mono-cropping, as different crops have varying susceptibility to pests and diseases. The economic viability of crop rotation is also evident in improved market access. Diversified crops can cater to diverse consumer demands, leading to increased market opportunities and potentially higher prices for the produce. Profitability of a crop rotation system is rooted in its ability to optimize resource utilization, reduce input costs, enhance soil health, and capitalize on market diversity. Adopting such a sustainable agricultural practice can result in long-term financial gains for farmers, promoting both economic and environmental sustainability in agriculture.

Keywords: *Crop Rotation, Environmental Sustainability, Fertility, Productivity, Profitability etc.*

Introduction

Crop rotation is a sustainable and time-tested agricultural practice that involves growing different crops in a sequential and systematic manner on the same piece of land. This method is designed to optimize the use of resources, enhance soil health, and improve overall agricultural productivity. One of the key aspects of crop rotation is its positive impact on the profitability of farming operations. The profitability of a crop rotation system is rooted in its ability to address various challenges faced by farmers while maximizing yields and minimizing input costs. One of the primary advantages lies in the efficient utilization of nutrients in the soil. Different crops have different nutrient requirements, and by rotating crops, farmers can prevent the depletion of specific nutrients and maintain well-balanced soil fertility. This helps to reduce the need for synthetic fertilizers, cutting down on input costs and enhancing overall profitability. Crop rotation also plays a crucial role in pest and disease management. Continuous cultivation of the same crop in a given area can lead to a build-up of pests and diseases that specifically target that crop. By rotating crops, farmers disrupt the life cycles of these pests and reduce the risk of infestations. This, in turn, decreases the reliance on chemical pesticides, leading to cost savings and contributing to higher profitability. Additionally, healthier plants resulting from a diverse crop rotation are more resilient to diseases, further safeguarding yields. The improvement of soil structure is another critical factor contributing to the profitability of crop rotation. Different crops have varying root structures and depths, which, when rotated, help break up compacted soil and enhance water infiltration. This reduces the risk of water logging and erosion, especially in areas with heavy rainfall. Improved soil structure also promotes better aeration and root development, creating an optimal environment for plant growth. As a result, crops grown in well-structured soils are more likely to thrive, leading to increased yields and, consequently, higher profits for farmers. Crop rotation provides an effective means of weed control. Weeds that are specific to certain crops can be suppressed by rotating with crops that are less susceptible to those particular weed species. This natural weed management strategy reduces the need for herbicides, lowering input costs and contributing to the economic sustainability of farming operations. In addition to these direct economic benefits, crop rotation can enhance the long-term sustainability of agriculture. The diversification of crops helps to mitigate the risks associated with market fluctuations and changing climatic conditions. Farmers relying on a single crop are more vulnerable to market price volatility and adverse weather events, whereas those practicing crop rotation are better positioned to adapt to such challenges. The profitability of a crop rotation system is multifaceted, encompassing nutrient management, pest and disease control, soil structure improvement, weed suppression, and overall risk mitigation. By adopting this sustainable agricultural practice, farmers not only achieve short-term economic gains but also contribute to the long-term health and resilience of their farming systems. As the global agricultural sector faces increasing pressure to produce more with fewer resources, crop rotation stands out as a viable and economically sound strategy for ensuring the continued success of farming operations.

Definition of Crop Rotation

Crop rotation is a systematic farming practice that involves growing different crops in sequential seasons on the same piece of land. This strategic alternation helps maintain soil health, reduce pest and disease incidence, and optimize nutrient levels. By breaking monoculture cycles, crop rotation enhances overall agricultural sustainability, mitigates risks, and often leads to increased crop yields.

Components of Crop Rotation

Components of crop rotation include a variety of crop selections for genetic variety and species richness, smart scheduling of rotations, and evaluation of complementing traits including growth behaviors and root systems. This responsive and adaptive system seeks to manage pests, improve soil health, maximize resources, and maintain agricultural output throughout time.

a. **Crop diversity and selection:** Crop diversity refers to the practice of cultivating a wide range of plant species and genetic variations within each species. It involves growing various crops such as cereals, legumes, vegetables, and fruits in a given area or over successive seasons. Genetic diversity is achieved by using different varieties or cultivars within each crop type. This agricultural strategy promotes resilience, adaptability, and sustainability in farming systems. By avoiding monoculture and incorporating diverse crops, farmers can enhance soil health, reduce pest and disease risks, improve nutrient cycling, and contribute to the conservation of plant genetic resources. Crop diversity is vital for building robust and sustainable agricultural ecosystems.

b. **Planning rotation schedules:** Creating rotation schedules, which involve planting crops on a field in a methodical manner across several seasons, is a crucial tactic in sustainable agriculture. When creating rotation plans, farmers carefully take into account variables including crop variety, soil health, and market demands. This method optimizes soil fertility, minimizes pest and disease burdens, and improves overall agricultural resilience. Rotation schedules are successful when they are regularly adjusted to changing environmental circumstances and meticulous record-keeping is maintained. This strategy helps to maintain the productivity and long-term health of farming systems by encouraging biodiversity and the effective use of resources, resulting in a more resilient and sustainable agricultural environment.

Soil Health and Fertility

Fertility and soil health are essential for productive agriculture. They include all of the physical, chemical, and biological characteristics of the soil that affect plant growth. Crop rotation and cover crops are examples of sustainable farming techniques that improve soil structure, microbial activity, and nutrient availability. To ensure ideal circumstances for plant growth, fertilizer management is guided by routine soil testing. Soil fertility and health are critical factors in sustainable farming techniques because they enhance crop yields, lessen their negative effects on the environment, and promote long-term agricultural sustainability.

a. **Nutrient cycling in crop rotation:** Crop rotation involves a dynamic process called nutrient cycling that is essential to preserving soil fertility. Since each crop has a different set of nutritional needs, crop rotation helps to slow down the loss of particular nutrients. Legumes, such as beans and peas, are essential because they work in symbiosis with nitrogen-fixing bacteria to fix atmospheric nitrogen into the soil. This natural enrichment minimizes the dependency on synthetic fertilizers. One crop's residue improves the structure and microbiological activity of the following crop by turning it into organic matter. The cyclical pattern of nutrient exchange promotes sustainable agriculture, makes the best use of resources, and enhances soil health, all of which support long-term productivity and environmental sustainability.

b. **Impact on soil organic matter:** Crop rotation encourages the accumulation and stability of soil organic matter, which has a substantial impact on it. A steady supply of organic wastes to the soil is ensured by including a variety of crops in the rotation cycle. This input improves soil structure, water retention, and nutrient availability by boosting microbial activity and encouraging the breakdown of organic matter. Increased soil organic matter also strengthens the soil's resilience to erosion and helps absorb carbon, thereby limiting the impact of climate change on agricultural ecosystems. Crop rotation promotes sustainable farming methods and healthier soils overall.

Pest and Disease Management

Pest and disease management in crop rotation is a strategic approach to minimize the impact of harmful organisms on agricultural yields. By disrupting pest and pathogen life cycles through thoughtful crop sequencing, farmers can reduce the risk of infestations.

a. **Disruption of pest and pathogen life cycles:** Disrupting pest and pathogen life cycles is a fundamental aspect of pest management in crop rotation. The practice involves strategically alternating crops to break the continuity of pests and pathogens, minimizing their buildup. By interrupting their life cycles, the population of harmful organisms is reduced, contributing to sustainable agriculture. This approach mitigates the need for excessive pesticide use, promoting a more balanced and environmentally friendly farming system that enhances crop health and overall productivity.

b. **Role of cover crops in pest control:** Cover crops play a crucial role in pest control within crop rotation systems. Acting as a natural ally against pests, cover crops provide habitat and food sources for beneficial insects that prey on harmful pests. This biological control helps disrupt pest life cycles, reducing their populations. Certain cover crops also release compounds that deter or inhibit pathogenic organisms, contributing to disease management. Furthermore, cover crops enhance overall biodiversity, fostering a balanced ecosystem. By integrating cover crops into crop rotation, farmers can achieve sustainable pest control, minimizing the reliance on chemical interventions, promoting soil health, and creating a resilient agricultural environment conducive to long-term productivity.

Yield Stability and Risk Mitigation

Yield stability and risk mitigation are critical aspects of crop rotation, ensuring consistent and reliable agricultural productivity. By diversifying the types of crops grown over successive seasons, crop rotation helps mitigate risks associated with factors like adverse weather conditions, pests, and diseases. This strategic approach minimizes the likelihood of total crop failure, as different crops respond differently to various challenges. Additionally, the practice spreads economic risks by offering a variety of income sources. Yield stability is enhanced through the optimization of soil health and nutrient balance achieved in well-planned crop rotations. Overall, crop rotation contributes to a resilient and sustainable farming system, promoting stability in yields and minimizing uncertainties.

a. **Diversification benefits:** The benefits of crop rotation for diversity are numerous and have a good effect on agricultural sustainability and the farm ecosystem. In the first place, it reduces risks by increasing exposure to pests, illnesses, and unfavorable weather, strengthening resilience to unanticipated difficulties, and decreasing the likelihood of crop failure. Because various crops react differently to environmental stressors, yield stability is improved, resulting in consistent productivity over seasons. By maximizing nutrient cycling and minimizing soil degradation, the technique fosters long-term fertility and improves soil health. Diverse crops interfere with the life cycles of pests and diseases, reducing the demand for chemical pesticides and promoting organic pest management. Crop diversification is a cost-effective strategy that satisfies consumer demand by offering a variety of revenue sources and mitigating market volatility. Crop rotation is further established as a key component of resilient and sustainable agriculture by the support of biodiversity, climate adaptation, better soil structure, efficient use of water, and alignment with sustainable practices.

b. **Reduction of yield risks due to weather and disease:** Crop rotation contributes significantly to the reduction of yield risks associated with weather and diseases. By diversifying crops, farmers minimize susceptibility to specific environmental conditions or pathogens that may affect a particular crop. Different crops respond uniquely to varying weather patterns and have diverse disease vulnerabilities. This strategic diversity ensures that even if one crop faces challenges, others in the rotation can thrive, leading to more stable overall yields. Additionally, the disruption of pest and pathogen life cycles through crop rotation minimizes the buildup of harmful organisms, reducing the risk of widespread disease outbreaks and enhancing the resilience of the entire agricultural system.

Economic Considerations

Economic considerations in crop rotation are pivotal for sustainable farming. Farmers assess market demand and profitability, optimizing crop selection for diversified income streams. Careful cost-benefit analyses guide resource allocation and input efficiency, minimizing financial risks. Government incentives, subsidies, and compliance with regulations are key factors. Long-term financial planning, adaptation to consumer preferences, and strategic investments in soil health contribute to economic resilience. Crop insurance and robust record-keeping further enhance risk management, ensuring a financially viable and sustainable agricultural operation.

Environmental Sustainability

Crop rotation is integral to environmental sustainability, offering numerous benefits. It enhances biodiversity, fostering a balanced ecosystem that supports beneficial insects, pollinators, and soil microorganisms. By disrupting pest and pathogen life cycles, crop rotation reduces the reliance on chemical pesticides, preserving natural ecosystems. Improved soil health through nutrient cycling and organic matter contributions mitigates soil degradation and erosion. This sustainable practice promotes efficient water use, as diverse crops have varying water requirements. Additionally, reduced dependence on synthetic fertilizers minimizes nutrient runoff, preserving water quality. Crop rotation aligns with climate-smart agriculture, contributing to carbon sequestration and adapting to changing environmental conditions. Overall, it establishes resilient and environmentally friendly farming systems.

Crop Selection and Rotation Planning

Crop selection and rotation planning are pivotal in sustainable crop rotation. Crop selection involves choosing crops based on market demand, profitability, and local adaptability. Diversification minimizes risk and optimizes resource use. Rotation planning dictates the sequential planting of crops, disrupting pest and disease life cycles while optimizing soil health. Careful consideration of economic factors, market trends, and climate conditions informs the order and duration of crops in rotation. A long-term vision for sustainability and adaptability, integration of technology, record-keeping, and risk mitigation through diversity and insurance contribute to the success of crop selection and rotation planning, ensuring resilient and productive agricultural practices.

Monitoring and Record-Keeping

Monitoring and record-keeping are pivotal in successful crop rotation strategies. Farmers systematically observe and document the sequence of crops planted in specific fields over time. Regular monitoring involves assessing soil health, crop performance, and pest prevalence. Detailed records include information on crop types, planting dates, and any associated practices such as cover cropping or fertilization.

Accurate documentation enables farmers to track the rotation cycle, helping prevent soil nutrient depletion and minimizing the risk of disease and pest buildup. This historical data informs decision-making for future planting seasons, allowing farmers to optimize yields and maintain sustainable agricultural practices. Digital tools and farm management software can enhance the efficiency of record-keeping, providing valuable insights for crop rotation planning. By prioritizing meticulous monitoring and record-keeping, farmers contribute to the longevity and productivity of their agricultural endeavors while promoting soil health and biodiversity.

Government Incentives and Subsidies

Governments incentivize crop rotation through financial support, offering subsidies, grants, and tax incentives. Farmers adopting sustainable crop rotation practices may receive direct payments or reduced taxes. Educational programs and research funding further promote awareness and understanding of the benefits. Some governments also provide subsidies specifically for cover crops, integral to effective crop rotation.

Case Studies and Success Stories

Several case studies and success stories highlight the positive impact of crop rotation on agricultural sustainability. One notable example is the Iowa State University's research, showcasing improved yields and reduced reliance on chemical inputs through diversified crop rotations. In Canada, farmers in the Prairie Provinces have successfully implemented canola-wheat rotations, breaking disease cycles and enhancing soil fertility.

Additionally, the Sustainable Agriculture Research and Education (SARE) program in the United States features numerous case studies demonstrating increased resilience and profitability through crop rotation practices, emphasizing the benefits for both the environment and farmers' economic outcomes. These cases underscore the practical advantages and success achieved by adopting thoughtful crop rotation strategies.

1. Real-world examples of successful crop rotation systems

A. **Three-Sister Farming (North America):** Indigenous communities in North America historically practiced the "Three Sisters" approach, combining corn, beans, and squash in a symbiotic system. Corn provides support for beans to climb, beans fix nitrogen in the soil, and squash acts as a ground cover, reducing weeds. This system enhances soil fertility, reduces pest pressure, and promotes sustainable agriculture.

B. **European Crop Rotation (Western Europe):** Traditional European crop rotations involve alternating cereal crops with legumes and fallow periods. This practice helps break pest and disease cycles, maintains soil fertility through nitrogen fixation, and prevents soil erosion. Farmers in regions like France and Germany have successfully employed this system for centuries.

C. **Canola-Wheat Rotation (Canada):** Canadian farmers, particularly in the Prairie Provinces, have implemented successful canola-wheat rotations. Canola's deep-rooted structure helps break compacted soil layers, improving water infiltration for subsequent wheat crops. This rotation has shown to reduce disease pressure on both crops and enhance overall yields and soil health.

D. **Cover Crop Integration (United States):** Farmers in the United States have adopted cover crops, such as legumes and grasses, in their rotations. Cover crops help prevent soil erosion, suppress weeds, and enhance soil structure. Successful integration has been observed in states like Iowa and Illinois, contributing to improved soil health and increased resilience against extreme weather events.

E. **Coffee-Banana Rotation (Central America):** In Central American countries like Costa Rica, coffee-banana rotations have proven successful. Coffee plants benefit from the shade provided by mature banana plants, reducing the need for synthetic inputs. Additionally, the banana plants contribute organic matter to the soil when they are chopped down after fruiting, creating a sustainable and mutually beneficial system.

Challenges and Considerations in Crop Rotation

Crop rotation faces challenges in managing diverse crops, adapting infrastructure, and responding to market dynamics. Transitions between crops require careful planning for soil preparation. The need for adaptable infrastructure impacts operational efficiency. Aligning rotations with market demands poses uncertainties, and effective pest and disease management is essential. Skill development for crop diversity management is crucial, along with implementing quality control measures. Climate variability adds complexity, requiring farmers to anticipate and mitigate potential impacts on crop growth for a resilient agricultural system.

a. **Potential drawbacks and challenges of crop rotation:** Crop rotation, while beneficial, presents potential drawbacks and challenges. Continuous rotations may lead to increased weed pressure, requiring additional management. Certain crops may be more susceptible to diseases or pests, impacting overall yields. Adaptation of machinery for diverse crops can be costly. Farmers may face market uncertainties and consumer demand fluctuations when planning rotations. Additionally, proper execution of rotations demands expertise and continual monitoring, adding complexity to farm management. Addressing these challenges is essential for a successful and sustainable crop rotation system.

b. **Strategies to overcome obstacles in crop rotation:** To overcome obstacles in crop rotation, farmers can employ several strategies. Implementing integrated pest management practices helps control pests and diseases. Diversifying crop selections within rotations minimizes susceptibility to specific issues. Utilizing cover crops enhances soil health and reduces weed pressure. Regular soil testing guides nutrient management, addressing deficiencies. Investing in adaptable equipment streamlines diverse crop management. Staying informed about market trends aids in aligning rotations with consumer demand. Continuous education and skill development empower farmers to navigate challenges effectively.

Future Trends and Innovations of Crop Rotation

Future trends in crop rotation may involve advanced technologies and sustainable innovations. Precision agriculture and data analytics could optimize rotation planning, tailoring it to specific field conditions. Climate-resilient crop varieties may become integral, addressing changing environmental patterns. Innovations in robotic farming may automate crop transitions. Additionally, regenerative agriculture practices may gain prominence, emphasizing soil health. Continued research and development are expected to drive transformative changes; ensuring crop rotation remains a key strategy for sustainable and productive agriculture in the future.

1. Emerging technologies in crop rotation

Emerging technologies are revolutionizing crop rotation practices in agriculture:

a. **Precision Agriculture:** Precision agriculture is an innovative farming approach that utilizes technology such as GPS, sensors, and data analytics to optimize crop production. By collecting and analyzing data on factors like soil conditions, weather patterns, and crop health, farmers can make informed decisions, leading to more efficient resource utilization, increased yields, and reduced environmental impact.

b. **Drones and Remote Sensing:** Drones, equipped with advanced sensors, enable remote sensing by collecting data from inaccessible or hazardous areas. They're pivotal in agriculture, environmental monitoring, and disaster response. Capable of capturing high-resolution imagery, thermal data, and more, drones enhance efficiency in various fields, offering valuable insights for decision-making, resource management, and scientific research.

c. **Machine Learning and Artificial Intelligence:** Machine Learning and Artificial Intelligence involve developing algorithms and models that enable computers to learn from data and perform tasks without explicit programming. Machine Learning focuses on creating systems that improve with experience, while Artificial Intelligence aims to build machines capable of intelligent behavior, including problem-solving and decision-making, resembling human cognitive functions.

d. **Smart Farm Equipment:** Advanced machinery can adapt to different crops, optimizing planting and harvesting processes, reducing the challenges associated with diverse rotations.

e. **Biotechnology and Genetic Engineering:** Biotechnology harnesses biological systems, organisms, or derivatives for practical applications. Genetic engineering manipulates an organism's genetic material to achieve desired traits or functions. These technologies find use in medicine, agriculture, and industry, offering solutions like gene therapy, genetically modified crops, and biopharmaceuticals. While they hold promise for advancements, ethical and safety considerations underscore ongoing debates in their implementation and regulation.

f. **Sensor Technologies for Soil Health:** Various sensor technologies play a crucial role in assessing soil health. Soil moisture sensors measure water content, aiding in irrigation management. Soil nutrient sensors assess nutrient levels, guiding fertilization strategies. Electrical conductivity sensors provide insights into soil salinity. Additionally, soil temperature sensors help monitor thermal conditions. These technologies enable farmers to make informed decisions, optimizing crop yield and sustainability while minimizing environmental impact.

2. Trends shaping the future of sustainable agriculture for crop rotation

The future of sustainable agriculture, particularly in the realm of crop rotation, is influenced by key trends that prioritize environmental health and resource efficiency. Precision agriculture, integrating technologies like sensors and data analytics, is shaping the way farmers plan and execute crop rotations, optimizing resource use and minimizing environmental impact. Agro ecology principles, such as diverse crop rotations and cover cropping, are gaining prominence, enhancing soil health and biodiversity. Advances in genetic engineering contribute to the development of crops with improved disease resistance and higher resilience to changing climates. Smart farming equipment facilitates efficient transitions between crops, supporting diverse rotation schemes. Collaborative efforts and knowledge-sharing platforms foster the widespread adoption of sustainable crop rotation practices, ensuring a collective commitment to address food security and environmental sustainability challenges in agriculture.

Conclusion

Profitability of a crop rotation system is evident through its multifaceted benefits. By diversifying crops over successive seasons, farmers enhance soil fertility, reduce pest and disease pressure, and optimize resource utilization. This sustainable practice minimizes input costs and fosters long-term agricultural resilience. Improved yields, coupled with decreased reliance on external inputs, contribute to overall financial gains. Additionally, crop rotation mitigates environmental impact, aligning with modern agricultural trends favoring eco-friendly practices. In summary, the adoption of a well-managed crop rotation system proves to be a judicious investment, fostering economic viability, environmental sustainability, and the overall resilience of farming enterprises

References

Berzsenyi, Z., Gyorffy, B. and Lap, D. 2000. Effect of crop rotation and fertilization on maize and wheat yields and yield stability in a long-term experiment. *European Journal of Agronomy*, *13*(2-3): 225-244.

Colbach, N. and Debaeke, P. 1998. Integrating crop management and crop rotation effects into models of weed population dynamics: a review. *Weed science*, *46*(6): 717-728.

Davis, A.S., Hill, J.D., Chase, C.A., Johanns, A.M. and Liebman, M. 2012. Increasing cropping system diversity balances productivity, profitability and environmental health.

dos Santos Canalli, L.B., da Costa, G.V., Volsi, B., Leocádio, A.L.M., Neves, C.S.V.J. and Telles, T.S. 2020. Production and profitability of crop rotation systems in southern Brazil. *Semina: Ciências Agrárias*, *41*(6): 2541-2554.

Higgs, R.L., Peterson, A.E. and Paulson, W.H. 1990. Crop rotations sustainable and profitable. *Journal of Soil and Water Conservation*, *45*(1): 68-70.

Moghaddam, H.A., Ramroudi, M., Koohkan, S.A., Fanaei, H.R. and Moghaddam, A.A. 2011. Effects of crop rotation systems and nitrogen levels on wheat yield, some soil properties and weed population. *International Journal of Agri Science*, *1*(3): 156-163.

Smith, E.G., Zentner, R.P., Campbell, C.A., Lemke, R. and Brandt, K. 2017. Long□term crop rotation effects on production, grain quality, profitability, and risk in the northern Great Plains. *Agronomy Journal*, *109*(3): 957-967.

Stanger, T.F., Lauer, J.G. and Chavas, J.P. 2008. The profitability and risk of long□term cropping systems featuring different rotations and nitrogen rates. *Agronomy Journal*, *100*(1): 105-113.

Volsi, B., Bordin, I., Higashi, G.E. and Telles, T.S. 2020. Economic profitability of crop rotation systems in the Caiuá sandstone area. *Ciência Rural*, *50*.

Volsi, B., Higashi, G.E., Bordin, I. and Telles, T.S. 2021. Production and profitability of diversified agricultural systems. *Anais da Academia Brasileira de Ciências*, *93*.

Volsi, B., Higashi, G.E., Bordin, I. and Telles, T.S. 2022. The diversification of species in crop rotation increases the profitability of grain production systems. *Scientific Reports*, *12*(1): 198-200.

10

Modern Prospective of the Ancient Grain: Millets

Ruchi Rani Gangwar[1], Rahul Kumar Rai[2] and Somya Tewari[3]

[1 &3] Department of Agricultural Economics, CoA, GBPUAT, Pantnagar Uttarakhand

[2]Department of Agricultural Economics, CoA, BUAT, Banda, Uttar Pradesh

Abstract

Government of India had proposed to United Nations for declaring 2023 as International Year of Millets (IYOM). The proposal of India was supported by 72 countries and United Nation's General Assembly (UNGA) declared 2023 as International Year of Millets on 5th March, 2021. Now, Government of India has decided to celebrate IYOM, 2023 to make it peoples' movement so that the Indian millets, recipes, value added products are accepted globally. India is the largest producer of millets in the world. Accounting for 20 % of global production and 80% of Asia's production. Millets provide food security to millions of households and contribute to the economic efficiency of farming In India, Important millet crops grown in India.

Keywords: *Millets, Nutritional Security, Future Prospective, Government Initiatives for Millets etc.*

Introduction

The history of millets is long and fascinating, dating back thousands of years. Millets are believed to have originated in Africa and Asia, where they were some of the earliest cultivated crops. Evidence suggests that millet cultivation began around 7000-5000 BCE in parts of Africa. Millets spread from Africa to Asia and later to Europe. Archaeological findings indicate the cultivation of millets in ancient China, India, and the Middle East. Millets played a crucial role in the diets of ancient civilizations. In China, foxtail millet and proso millet were cultivated as early as the Neolithic period. In India, millets such as finger millet (ragi), pearl millet (bajra), and sorghum (jowar) have been traditional crops for thousands of years.

Millets continued to be important crops during medieval times, especially in regions with challenging climates where they thrived. In Europe, millets were commonly consumed during the medieval and Renaissance periods, but their popularity diminished with the introduction of other grains like wheat and barley. The colonial period brought changes in agricultural practices, and millets saw a decline in cultivation as other crops gained prominence. In many regions, millets were considered "poor man's food," leading to a decrease in their popularity. In recent decades, there has been a resurgence of interest in millets due to their nutritional benefits, resilience to drought, and suitability for sustainable agriculture. Organizations and governments have promoted millet cultivation as part of efforts to enhance food security, combat malnutrition, and promote sustainable farming practices. Millets have gained global recognition as important crops for food security, especially in the face of climate change. Efforts to promote millets as a gluten-free and nutritionally rich alternative have led to increased consumption in various parts of the world. Today, millets are recognized not only for their historical significance but also for their potential to address contemporary challenges in agriculture, nutrition, and sustainability.

India's Millets Scenario

India is the top millet producing country in the world. It contribute 42% of global millet production followed by Niger (10%), China(9%), Nigeria (6%), Mali (6%).The area under cultivation of millets has ranged from 12.29 to 15.48 million hectares from 2013-14 to 2021-22.Six states namely Rajasthan, Karnataka, Maharashtra, Uttar Pradesh, Haryana, and Gujarat accounts for more than 83 per cent share in total millet production. Rajasthan contributes 28.61 per cent of the total millet production in India. Multiple varieties of millets are produced in India such as Pearl Millets, Sorghum, Finger Millet, Foxtail, Kodo, Barnyard, Proso, Little Millet and Pseudo Millets like Buckwheat and Amaranths. Pearl millet (Bajra), Sorghum (Jowar) and Finger Millet (Ragi) constitutes the largest share in India's total production of millets.

Top 10 Largest Millets Producing States in India 2022

Rank wise State position	Share of Total Millet Production (%)	Major Millet Crops
Rajasthan	28	Bajra, Pearl Millet
Karnataka	18	Ragi, Finger Millet
Maharashtra	14	Jowar, Sorghum
Uttar Pradesh	12	Pearl Millet, Bajra
Gujarat	7	Bajra, Pearl Millet
Madhya Pradesh	6	Kodo Millet, Jowar
Tamil Nadu	5	Jowar, Little Millet
Andhra Pradesh	4	Jowar, Pearl Millet
Uttarakhand	2	Barnyard Millet, Foxtail Millet
Haryana	1	Bajra, Pearl Millet

Source: https://apeda.gov.in/milletportal/files/Statewise_Millet_Production.pdf

Nutritional Importance

Millets are pretty impressive little grains. They pack a nutritional punch that's hard to ignore. These tiny powerhouses are rich in nutrients like fiber, vitamins, and minerals. They're also gluten-free, making them a great option for those with gluten sensitivities or celiac disease. From a health perspective, millets offer benefits like improved digestion, better heart health, and stable blood sugar levels. Plus, they're a fantastic source of antioxidants, which help fight inflammation and oxidative stress in the body. Millets are also champions of sustainability. They're hardy crops that can thrive in diverse climates and require less water compared to some other grains. Growing millets can contribute to more resilient and sustainable agricultural practices. Let's not forget the cultural significance of millets. They've been staple foods in many regions for centuries, contributing to the culinary diversity of various cuisines around the world.

So, whether you're looking for a nutritious addition to your diet, a sustainable crop option, or a way to diversify your meals, millets definitely deserves a spot on your plate. Millets are like a nutritional goldmine. Their composition may vary slightly depending on the specific type of millet, but in general, they share some key characteristics. Here's a breakdown of their nutritional goodness:

- **Protein:** Millets are a good source of plant-based protein, which is essential for muscle repair, immune function, and overall growth.
- **Fiber:** High fiber content makes millets excellent for digestion and can help regulate blood sugar levels. It also contributes to a feeling of fullness, aiding in weight management.

- **Vitamins:** Millets contain various vitamins such as B-complex vitamins (like niacin, thiamine, and riboflavin), which play a crucial role in energy metabolism and overall well-being.
- **Minerals**: Millets are rich in minerals like iron, magnesium, phosphorus, and zinc. These minerals are essential for bone health, immune function, and maintaining overall vitality.
- **Antioxidants:** Millets contain antioxidants that help combat oxidative stress and inflammation in the body, contributing to long-term health.
- **Low Glycemic Index**: Millets generally have a low glycemic index, meaning they have a smaller impact on blood sugar levels compared to some other grains. This is especially beneficial for individuals managing diabetes.
- **Gluten-Free:** Millets are naturally gluten-free, making them a safe and nutritious choice for those with gluten sensitivities or celiac disease.

Including a variety of millets in your diet can provide a well-rounded mix of these nutrients, contributing to a balanced and wholesome approach to nutrition.

Types of Millets

There's quite a variety of millets out there, each with its unique flavor, texture, and nutritional profile and different growing season. The cropping seasons of millets can vary based on the specific type of millet and the region in which they are grown. However, here's a general overview:

- **Pearl Millet (Bajra):** Widely grown in Africa and the Indian subcontinent, pearl millet is known for its high protein and iron content. It's often used in flatbreads and porridge. Typically, pearl millet is grown during the summer season. It is a warm-season crop that thrives in high temperatures.
- **Finger Millet (Ragi):** This millet is rich in calcium, making it excellent for bone health. It's a staple in parts of Africa and Asia, often used to make porridge, flatbreads, or even fermented beverages. Ragi is adaptable to various climates, but it is often grown as a rain-fed crop during the monsoon season. In some regions, it can also be cultivated during the post-monsoon or winter season.

- **Foxtail Millet (Kangni):** A popular millet in East Asia, foxtail millet is rich in fiber and low in glycemic index. It's commonly used in porridge and rice dishes. Foxtail millet is a warm-season crop, and it is usually grown during the summer. It is well-suited for regions with high temperatures.

- **Sorghum (Jowar):** Sorghum is versatile and used in various forms, including whole grains, flour, or syrup. It's gluten-free and rich in antioxidants. Sorghum is a versatile crop that can be grown in both kharif (rainy) and rabi (winter) seasons, depending on the region. In some areas, it is cultivated during the summer as well.

- **Little Millet (Kutki)**: Packed with nutrients, little millet is often consumed in India. It's versatile and can be used in various dishes, from porridge to pulao. Little millet is a summer crop, and it is often grown during the hot and dry seasons.

- **Proso Millet (Common Millet):** This millet is cultivated in many parts of the world and is known for its drought-resistant nature. It's often used in birdseed but is gaining popularity as a human food due to its nutritional benefits. Proso millet is a warm-season crop, and it is typically grown during the summer months.

- **Barnyard Millet (Sanwa):** Commonly consumed in parts of India, barnyard millet is rich in fiber and has a nutty flavor. It's used in various dishes like porridge, upma, and dosas. Barnyard millet is a fast-growing crop that is often cultivated during the kharif (rainy) season. It is also grown as a Rabi (winter) crop in some regions.
- **Kodo Millet (Kodra):** Native to India, kodo millet is rich in dietary fiber and has a low glycemic index. It's often used in traditional Indian dishes. Kodo millet is primarily a rain-fed crop and is typically sown during the rainy season.

It's important to note that the specific cropping seasons can vary based on factors such as climate, soil conditions, and regional agricultural practices. Farmers often adapt their cultivation practices based on the local climate patterns and the characteristics of each millet variety.

Incorporating a variety of millets into your diet not only adds diversity to your meals but also ensures a range of nutritional benefits. Each type brings its own set of nutrients to the table.

Millets are like a treasure trove of health benefits. Here are some of the fantastic advantages they bring to the table:

- **Nutrient-Rich:** Millets are packed with essential nutrients, including protein, fiber, vitamins, and minerals. They provide a well-rounded nutritional profile that supports overall health. The protein content of millets ranges between 10% and 12% while the protein content in foxtail millet (12.3%) is higher than that of wheat and rice.
- **Heart Health**: The fiber and magnesium content in millets contribute to heart health. They can help lower cholesterol levels, regulate blood pressure, and reduce the risk of cardiovascular diseases.
- **Digestive Health:** The high fiber content in millets aids digestion, prevents constipation, and promotes a healthy digestive system. It also helps in maintaining a stable blood sugar level. The dietary fibre (10–12%) content of millets is also higher compared to some of the staple cereals.
- **Gluten-Free Goodness**: Millets are naturally gluten-free, making them a fantastic option for those with gluten sensitivities or celiac disease. They provide a safe alternative for individuals who need to avoid gluten.

- **Weight Management**: The combination of fiber and protein in millets can contribute to a feeling of fullness, helping with weight management by reducing overall calorie intake.
- **Bone Health:** Millets such as finger millet (ragi) are rich in calcium, supporting bone health and preventing conditions like osteoporosis.
- **Antioxidant Power**: Millets contain antioxidants that help fight oxidative stress and inflammation in the body, potentially reducing the risk of chronic diseases.
- **Diabetes Management**: Millets have a low glycemic index, meaning they cause a slower rise in blood sugar levels. This can be beneficial for individuals managing diabetes, helping to stabilize blood sugar.
- **Sustainable Agriculture**: Millets are hardy crops that can thrive in diverse climates with minimal water requirements. Cultivating millets contributes to more sustainable and resilient agricultural practices.
- **Culinary Diversity**: Millets add variety to your diet, offering different textures and flavors. They can be used in a wide range of dishes, from porridge and flatbreads to salads and pilafs.

Future Prospective

The future prospects of millets appear promising, with growing interest and recognition for their numerous benefits. Here are some potential aspects of the future of millets:

- **Nutritional Awareness**: As people become more health-conscious, there is likely to be a continued increase in the awareness of the nutritional benefits of millets. Millets, being rich in fiber, protein, and various vitamins and minerals, may become more popular among those seeking nutritious and diverse food options.
- **Global Demand and Trade**: The global demand for millets is expected to rise, driven by factors such as the increasing popularity of gluten-free diets, a growing interest in ancient grains, and a focus on sustainable and resilient agriculture. This could lead to an expansion of millet cultivation and trade on a global scale.
- **Climate Resilience**: Millets are known for their resilience to challenging environmental conditions, including drought and high temperatures. As climate change continues to pose challenges to traditional agriculture, the cultivation of millets may become more widespread as a resilient and adaptable crop.

- **Research and Innovation:** Ongoing research and innovation in agriculture may lead to the development of improved millet varieties with enhanced nutritional content, better yield, and resistance to pests and diseases. This can contribute to the overall sustainability and productivity of millet farming.
- **Policy Support**: Governments and agricultural organizations may implement policies to encourage millet cultivation, recognizing their importance in achieving food security, promoting sustainable agriculture, and addressing malnutrition.
- **Diversification of Food Products**: Millets may find their way into an increasing variety of food products, including baked goods, snacks, cereals, and beverages. The versatility of millets makes them suitable for diverse culinary applications.
- **Culinary Integration:** Millets may become more integrated into mainstream diets, both in traditional millet-consuming regions and in areas where they are less commonly consumed. Culinary innovation and the promotion of millet-based recipes can play a role in this integration.
- **Community Empowerment**: Millet cultivation can empower local communities, especially in regions with challenging agricultural conditions. Small-scale farmers may benefit from the cultivation of millets, contributing to rural livelihoods and economic development.

While these are potential trends, the future of millets will depend on various factors, including global food trends, research and development initiatives, policy support, and consumer preferences. The versatility, nutritional value, and adaptability of millets position them as important crops for the sustainable future of agriculture and food security.

Government Initiatives for Promoting Millets

- The Year 2023 is being celebrated as International Year of Millets-2023 for promotion of millets.
- Millet Awareness Quizzes/Competition and Conferences are being conducted by Food Corporation of India and Central Warehousing Corporation of Department of Food and Public Distribution.
- Millets has been termed as 'Nutri-cereals' owing to their health benefits. All the offices/Central Public Sector Enterprises of Department of Food and Public Distribution have been directed to introduce and promote millets in their canteens.

- Government of India fixes Minimum Support Prices (MSPs) for twenty-two (22) mandated crops based on the recommendations of the Commission for Agricultural Costs & Prices (CACP) after considering the views of concerned State Governments and Central Ministries/ Departments. The millets such as Ragi, Jowar and Bajra are covered under Minimum Support Price (MSP).
- Government makes allocation of coarse grains (millets) under Public Distribution System (PDS) and Other Welfare Schemes (OWSs) including Wheat Based Nutrition Programme (WBNP) and Pradhan Mantri Poshan Shakti Nirman (PM-POSHAN) Schemes based on the requirement projected by the State Govts., Ministry of Women and Child Development and Department of School Education & Literacy subject to the availability of coarse grains (millets).
- State Governments are also being requested repeatedly through written communications and in various meetings convened with them to procure millets and distribute under Targeted Public Distribution System (TPDS), Pradhan Mantri Poshan Shakti Nirman (PM-POSHAN), Integrated Child Development Services (ICDS) and Other Welfare Scheme (OWS).
- To promote the consumption of millets/coarse grains, distribution period of these commodities has been enhanced to 6-10 months from earlier period 3 months. Further provisions of inter-state transportation and advance subsidy have been incorporated in the guidelines.

Conclusion

Millets are a type of grain that provides various health benefits to the body by containing vitamins and minerals. Also, it is rich in dietary fibre, which helps to keep the digestive system healthy. It can also be used as a good substitute for rice and wheat. It is considered as one of the best grains for weight loss in India. Initially, millets were thought to be inferior to other cereals such as wheat or rice because they contain less gluten and are considered easy to digest. But some studies show that millets are beneficial as they contain essential nutrients such as proteins, amino acids, insoluble fibre etc that leads to better health and weight loss. This new demand for millets, leading to higher prices, can make their cultivation profitable, ensuring the legitimate place for millets in the national food basket.

References

Annual Reports, Ministry of Agriculture & Farmers Welfare, Department of Agriculture, Cooperation & Farmers Welfare, Govt. of India, 2022.

Deshpande, R.S., 2004. Coarse cereals in a drought-prone region: A study in Karnataka (Vol. 2). Institute for Social and Economic Change.

GoI, 2010-11: Economic Survey, Ministry of Finance, New Delhi.

https://apeda.gov.in/milletportal/files/Indian_Superfood_Millet_APEDA_Report.pdf

https://idronline.org/article/agriculture/millet-cultivation-history-and-trends/

https://pib.gov.in/PressReleasePage.aspx?PRID=1907194

ICAR-Indian Institute of Millets Research, Hyderabad website. https://millets.res.in

Kane-Potaka, J., Anitha, S., Tsusaka, T.W., Botha, R., Budumuru, M., Upadhyay, S., Kumar, P., Mallesh, K., Hunasgi, R., Jalagam, A.K. and Nedumaran, S., 2021. Assessing millets and sorghum consumption behavior in urban India: A large-scale survey. Frontiers in sustainable food systems, 5: 680-777.

Rao, B.D., Bhandari, R. and Tonapi, V.A., 2021. White Paper on Millets–A Policy Note on Mainstreaming Millets for Nutrition Security. ICAR-Indian Institute of Millets Research (IIMR), Rajendranagar, Hyderabad.

Rao, B.D., Reddy, C.S. and Seetharama, N., 2007. Reorientation of Investment in R&D of Millets for Food Security— the Case of Sorghum in India. Agricultural Situation in India, 64: 303-305.

Vision-Indian Institute of Millets Research, Hyderabad, 2025.

11

Reform of Agricultural Market by e-NAM and It's Role in Agriculture Sector in Madhya Pradesh

Rajendra Singh Bareliya[1] and Deepak Rathi[2] and Rahul Kumar Rai[3]

[1&2]Agro-Economic Research Centre, JNKVV, Jabalpur, Madhya Pradesh
[3]Department of Agriculture Economic, CoA, BUAT, Banda, Uttar Pradesh

Abstracts

The National Agriculture Market (e-NAM) is a pan-India electronic trading portal which networks the existing APMC mandis to create a unified national market for agricultural commodities. National Information Centre (NIC) for providing necessary servers to host e-NAM portal, state governments and marketing boards providing storage and warehousing facilities, regulate and dispute resolution mechanism to APMC markets, and APMC markets for implementation of physical and online trading. The top ten state have been covered 83.00 percent e NAM APMC and other state have been covered only 17.00 per cent APMC in India. There are about 2477 principal regulated markets based on location (the APMCs) and 4843 sub-market yards regulated by the respective APMCs in India. Trading transparency through better price discovery and e-NAM Live Price Information, Access to more markets and buyers. The link to logistics will be provided on e-NAM that can be accessed to avail the transportation services.

Keywords: *Reforms, e-NAM, Agriculture, Market, Madhya Pradesh, etc.*

Introduction

India is predominantly an agrarian economy, with agricultural sector engaging about half of the workforce. According to a survey conducted by NABARD in 2016-17, about 48% households in India are agricultural households (Bisena and Kumar, 2018). The Emerging changes in agriculture marketing environment of the country i.e. electronic market, model act, warehousing, pledge loan, contract farming, etc. are ushering in opportunities for new formats of markets which are effective in responding to demand and supply (Gupta and Badal, 2018). The National Agriculture Market (e NAM) is a pan-India electronic trading portal which networks the existing APMC mandis to create a unified national market for agricultural commodities. The Portal is managed by Small Farmers' Agribusiness Consortium (SFAC) with the technology provider, NFCL's iKisan division. A similar project was initiated by the Congress government in the State of Karnataka, during UPA tenure and had been a great success. NDA government has rolled it out nationally (Nedumaran and Manida, 2019). The Small Farmers Agribusiness Consortium (SFAC) is the lead agency for implementing e NAM under the aegis of Ministry of Agriculture and Farmers' Welfare, Government of India. Launched by Prime Minister Narendra Modi on April 14, 2016, e-NAM is completely funded by the Central Government. The e-NAM platform promotes better marketing opportunities for the farmers to sell their produce through online competitive and transparent price discovery system and online payment facility and the portal provides single window services for all APMC related information and services. This includes commodity arrivals, quality & prices, buy & sell offers and e-payment settlement directly into farmers' account, among other services. It aims to promote uniformity in agriculture marketing by streamlining of procedures across the integrated markets, removing information asymmetry between buyers and sellers and promoting real-time price discovery based on actual demand and supply. The scheme envisages deployment of a common e-market platform of 585 selected regulated wholesale agriculture market yards by March, 2018. The common electronic trading portal will be called as e-NAM (https://enam.gov.in). This abolishes the various physical managements at multiple levels and with multiple market fees.

It is a program of liking markets with huge effort of many agencies involved like the central government which providing uniform policy framework, Small Farmers Agribusiness Consortium (SFAC) playing the role of lead agency along with strategic private partner Nagarjuna Fertilizers and Chemicals Limited for maintenance of the portal, Directorate of Marketing and Inspection (DMI) for providing technical support for harmonization of standards for commodities and assaying facilities, National Information Centre (NIC) for providing necessary servers to host e-NAM portal, state governments and marketing boards providing storage and warehousing facilities, regulate and dispute resolution mechanism to APMC markets and APMC markets for implementation of physical and online trading (Jatana and Goswami, 2021). It is a virtual market connecting the current APMC Mandis electronically with a subject of "one country, one market." e-NAM advances consistency, smoothing out of systems over the incorporated markets, evacuates the data hole among purchasers and dealers, and advances ongoing value disclosure dependent on genuine interest and flexibly in the market (Singh *et al*,2021). Although APMC markets were established with the good intention of facilitating fair market practices by introducing open outcry system of auctioning, they have become breeding grounds to exploit the farmers over the years (Reddy and Mehjabeen, 2019).

Present status of e-NAM in Madhya Pradesh

The government have been total numbers of e NAM APMC 1388 at present in India out of which the highest number of e NAM APMC 11.31 percent of Tamil Nadu followed by Rajasthan, Gujarat, Madhya Pradesh, Maharashtra, Uttar Pradesh, Haryana Punjab, Odisha and Telangana i.e., 10.45, 10.37, 10.01,9.51, 9.01, 7.78, 5.69, 4.76, and 4.11 per cent respectively. The Madhya Pradesh has fourth position with 139 APMC Registered in e NAM.

Table 1: Top ten state wise registered on e-NAM mandis in India

S. No.	Name of State/UT	Mandis registered on e-NAM	Percent (%)
1	Tamil Nadu	157.00	11.31
2	Rajasthan	145.00	10.45
3	Gujarat	144.00	10.37
4	Madhya Pradesh	139.00	10.01
5	Maharashtra	132.00	9.51
6	Uttar Pradesh	125.00	9.01
7	Haryana	108.00	7.78
8	Punjab	79.00	5.69
9	Odisha	66.00	4.76
10	Telangana	57.00	4.11
11	Others	236.00	17.00
12	India	1388.00	100.00

Source: enam.gov.in

The top ten state have been covered 83.00 per cent e NAM, mandi and other state have been covered only 17.00 per cent APMC in India (Table 01). There are about 2477 principal regulated markets based on location (the APMCs) and 4843 sub-market yards regulated by the respective APMCs in India (www.google.com).

Table: 2: The Mandis, traders and unified licenses in India and Madhya Pradesh

Particulars	**India**	**Madhya Pradesh (%)**
Mandis registered on e-NAM	1389	139 (10.01)
Registered Traders on e-NAM	250915	22639 (9.01)
No. of Unified licenses issued by State	169527	1074 (0.63)

Source: enam.gov.in, Parenthesis show the per cent of India

The India have been found to be total number 1389 mandi registered on, e NAM, out of which mandi found 139 (10.00%) in Madhya Pradesh state of total. Whereas the total numbers of Registered Traders 250915 in India out of which 22639 (9.01%) traders were found to be in Madhya Pradesh. The Number of Unified licenses issued 169527 by State government in the country out of which Madhya Pradesh share only 0.63 per cent only. That position of e NAM licenses in the state to be worrying at present (Table 02).

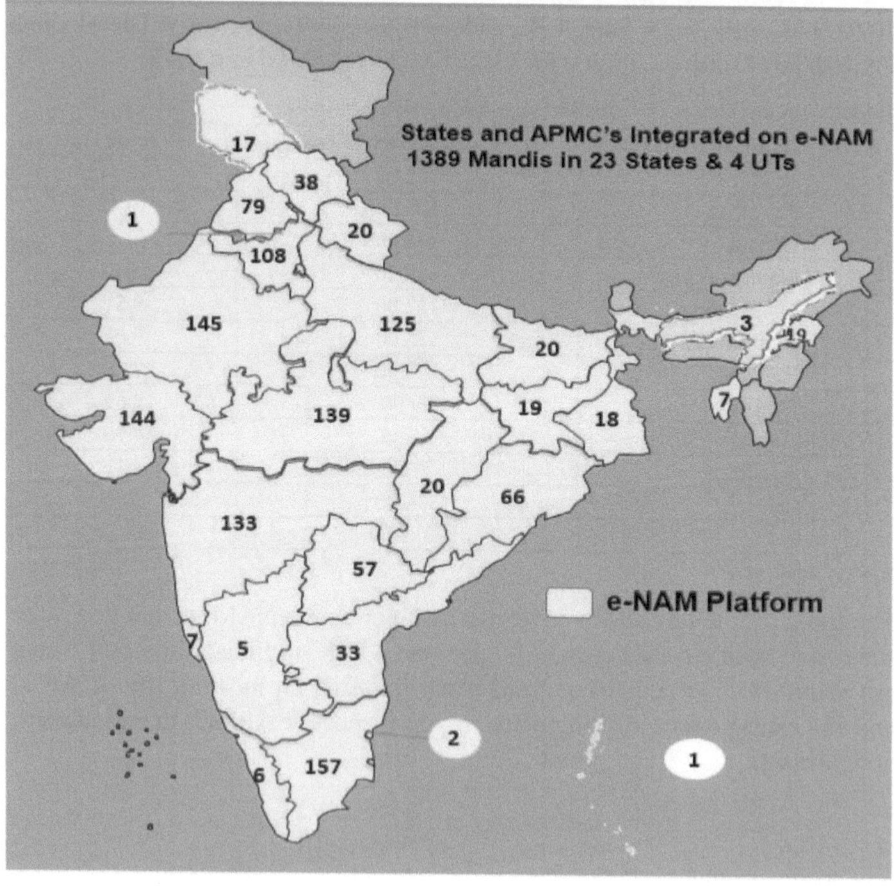

Source: e-NAM official website

Role of e-NAM

In May 2020, the Government of India had announced the integration of 38 new Mandis with e-National Agriculture Market (e-NAM). With the addition of these 38 Mandis the total tally of Mandis reached the milestone of 1000 spanning across 18 states and 3 Union Territories. Trading transparency through better price discovery and e-NAM Live Price Information, Access to more markets and buyers. Real-time information on prices and arrivals by mandis. Instant Payments - Will be able to build a healthy financial profile. e-NAM facilitates payment using online modes like RTGS, NEFT and BHIM UPI, with instant payment into the bank account of the respective sellers. Farmer get higher prices of commodity as compared to others market. Farmer Producer Organization (FPO) directly and easy connected to e NAM for commodity selling.

1. **SFAC**: Small Farmers' Agribusiness Consortium (SFAC) is the lead promoter of National Agricultural Market (e NAM). SFAC is formulated under the Department of Agriculture, Cooperation & Farmers' Welfare (DAC&FW). SFAC through open tender selects a Partner to develop, operate, and maintain the NAM e-platform.

2. **Registration Guidelines:** Registration is the process for registration of farmers/ traders/ FPOs/ Mandi board and logistic providers in the e NAM platform. For connect to each other's for marketing process.

   a) **For Farmers/ Traders:** The registration of farmers may be done through http://www.enam.gov.in website select the registration types farmer or traders and select the desired "APMC" thereafter uploaded your passport size photograph & correct Email ID to receive Login ID and Password in same. Once successfully registered you will receive a Temporary Login ID & Password in the given e-mail. Login to the Dashboard by clicking icon on www.enam.gov.in/web through the system. User will find a Flashing Message on the Dashboard as: "Click here to register with APMC". Click on the Flashing Link which will redirect you to Registration Page for filling/updating details. It will be sent for Approval to your selected APMC after KYC is completed. After Successful Login to your Dash board you will be able to see all APMC address details. After Successful Submission user will receive an e-mail confirming the submission of application to concerned APMC with status of the Application as Submitted/In progress—approved—rejected. Then you can contact to your respective Mandi/ APMC for the same.

b) **For FPOs/FPCs:** The FPOs/FPCs can register on e-NAM Portal via website (www.enam.gov.in) or mobile app or providing following details at nearest e-NAM mandi: - Name of FPOs/ FPCs- Name, address, email Id and contact no. of authorized person (MD/CEO / Manager)- Bank account Details (Name of Bank, Branch, Account no. IFSC Code)

c) **For Mandi Board:** The State Agriculture Marketing Boards interested to integrate their mandis with NAM are required to carry out following reforms in their APMC Act.

a) Single trading license (Unified) to be valid across the state

b) Single point levy of market fee across the state; and

c) Provision for e-auction/ e-trading as a mode of price discovery

d) **For logistics Provider:** The e-NAM has launched Logistics module on its platform to support its users with their supply chain requirements both for inbound (farm to APMC) and outbound (APMC to destination). The link to logistics will be provided on e-NAM that can be accessed to avail the transportation services.

3. Commodity traded by e NAM

The e NAM can be 209 commodities marketed by this platform out of vegetables (57), fruits (43), food grain (33), spices (16) oil seed (14), and miscellaneous (46) under the e NAM platform. The vegetables commodities were marketed in large quantity followed by fruits, food grains, spices, oilseed and others respectively.

4. e-NAM Process Flow

1) **Entry the Gate:** Marketed process flow starts from the Gate entry. First of a farmer provide the information about name, name of commodity, village, quantity of commodity, mobile numbers, address etc. to mandi officers.

2) **Unloading of Commodity at Auction:** The farmers after entry of gate unloading the commodity at platform for e bidding process. But before bidding process mandi staff collect the commodity sample for quality purpose about 250-500 gm and provide the sample reports to farmers.

Fig. Shows the e NAM marketed process

3) **Generate e-bidding:** The mandi officers generate e bidding based on sample report and decide the maximum bidding time. Traders will be quoting their price electronically for their interested lot. That time e bidding show on the display board at APMC area.

4) **Bid Declaration:** The commodity sold out or not that will be decided by farmers after once the bidding time is over, farmers get the information about maximum price and commodity lot number on his registered mobile number. if the farmers have satisfied to the price, then sold if not, he was wait for next bidding process.

5) **Weighment of sold commodity:** After successful completion of the auction process, the next process is Weighment of the commodity. Weighing of the commodity is done on the electronic weighing machine by the weigh men appointed by the mandi.

6) **Generation of Sale Agreement:** The mandis officer's primary bill will be generated after weighment. Description about trader's name, licences number, farmer's details, agreement number, commodity detail, packaging type and weight of bag, different types of fees and date etc.

7) **Payment to farmers:** After bill generate traders will be pay the amount to farmers by online mode as RTGS/NEFT in the farmers account and some amount pay cash that farmers may be pay to vehicle rent etc.

8) **Gate exit pass and gate exit:** The completion of traded and payment process successfully then after mandis official generate the gate exit pass to farmers which contains gate exit number, exit type, Vehicle number, APMC detail, Trader, lot type, Commodity, lot code, Bag type, number of bag and weight/ total number (Kumar and Pant, 2020).

Conclusion

The National Agriculture Market (e-NAM) is a pan-India electronic trading portal which networks the existing APMC mandis to create a unified national market for agricultural commodities. The e-NAM portal provides single window services for all APMC related information and services. The scheme envisages deployment of a common e-market platform of 585 selected[1] regulated wholesale agriculture market yards by March, 2018. The top ten state have been covered 83.00 percent e-NAM APMC and other state have been covered only 17.00 per cent APMC in India. Trading transparency through better price discovery and e-NAM Live Price Information, Access to more markets and buyers. Real-time information on prices and arrivals by mandis. Instant Payments - Will be able to build a healthy financial profile. e-NAM facilitates payment using online modes like RTGS, NEFT and BHIM UPI, with instant payment into the bank account of the respective sellers It will be sent for Approval to your selected APMC after KYC is completed. After Successful Login to your Dash board you will be able to see all APMC address details. Traders will be quoting their price electronically for their interested lot. That time e bidding show on the display board at APMC area.

References

Bisen, J. and Kumar, R., 2018. Agricultural marketing reforms and e-national agricultural market (e-NAM) in India: A review. Agricultural Economics Research Review, 31(conf): 167-176.

Gupta, S. and Badal, P.S., 2018. E-national Agricultural Market (e-NAM) in India: A Review. BHU Management Review, 6(1): 48-57.

Goswami, M. and Jatana, R., 2021. An analytical study on the functioning of eNAM (with special reference to rajasthan). International Journal of Research Culture Society, 5.

Kumar, S.A.D. and Pant, S.C., 2020. Benefits of e-NAM Process to Farmers–A Study. Internet: https://ccsniam. gov. in/images/pdfs/Benefit-of-eNAM-process-to-Farmer-A-Study.

Nedumaran, G., 2019. Trends and Impacts of e-NAM in India.

Reddy, A.A. and Mehjabeen, 2019. Electronic national agricultural markets, impacts, problems and way forward. IIM Kozhikode Society & Management Review, 8(2): 143-155.

Singh, D.P., Patil, C. and Meena, S.S., 2021. Impact of online agriculture marketing policy–e-NAM (electronic National Agriculture Market) on prices and arrivals of agricultural commodities in Punjab, India. Int. J. Curr. Microbiol. App. Sci, 10(02): 1573-1582.

Important links: https://enam.gov.in.

12

Soil Salinity and its Reclamation

***Shivam Singh*[1]*, Ankit Tiwari*[2]*, Himanshu Panday*[3] *and Mahendra Pratap Singh*[4]**

[1]*Department of Soil Science & Agricultural Chemistry, CoA, SVBPUAT Meerut, Uttar Pradesh*

[2]*Department of Agronomy, CoA, CSAUAT, Kanpur, Uttar Pradesh*

[3]*YP-II, ICAR-Indian Institute of Sugarcane Research, Lucknow, Uttar Pradesh*

[4]*Krishi Vigayn Kendra, Sonbhadra, ANDUAT, Kumarganj, Ayodhya Uttar Pradesh*

Abstract

The saline soil contains sufficient neutral soluble salts to adversely affect majority of crop plants. Saline soils are measured by electrical conductivity of more than 4 dS/m at 25 °C on the basis of salt concentration. The various methods of saline soil reclamation were used, physical methods including scraping, flushing, and leaching, agronomic practices, crop rotation, selective crop cultivation and various irrigation techniques, soil drainage, application of gypsum, and also biological methods of saline soil including applying manures, green manures, and bulky organic manures etc. Salinization is a salt condition in which accumulation of salt at the high concentration level that are responsible for reduced to plants growth and developments and pH also high. The minimum pH of soil should be 6.5 to 7.5, Plants will be healthy. Some plants are tolerant that are able in saline condition that is called salt tolerance crop plants like Barley (cereal crop) and others crops given in this chapter. The growth of plants depends on pH value, so before the sowing of crops check the pH of soil and check also salt concentration by Electric Conductivity (EC) for the high crop production and productivity.

Keywords: *Soil Salinity, Reclamation Methods, pH Value, Tolerance Crops*

Introduction

Salinity is a form of salt that occurs naturally within soils or water. When soil becoming salty form that is called salinization, it can be caused by mineral weathering called as natural process but it can be caused by irrigation called as artificial process. Saline or salty soil have a high salt content like sodium chloride (NaCl also called table salt). Salinity is like- abiotic stress for plants and reduced the plant growth but not for all because some of them crops are survive in better way for growth and development such as barley, sugerbeet, cotton and mustard etc. Basically, stresses are two types-biotic and abiotic stress. Abiotic stress are also called nonliving stress like- drought, salinity, submergence, heat, cold and temperature that are affect to plant life cycle so that are called as abiotic stress. Natural water and all types of soil both contain soluble salts. Amount of salt affected to weather of soil, when excessive salt present in soil and it adversely affects to crop growth and developments. Salt affected soils exist under arid and semi-arid climates. Major constitutions of soluble salt in soil are the cations and anions. Cations are contain sodium, calcium, magnesium and anions are contains chloride, sulphate, carbonate and bicarbonate. Saline soil have <8.5 pH, >4EC (dS/m) and <15 ESP. where EC is Eclectic conductivity that is measure of soil salinity and ESP stand for Exchangeable sodium percentage.

1. Causes of Soil Salinity

There is mainly two causes made salty condition of soil: Primary and Secondary

Primary causes of soil Salinity: Primary causes of salinization is are manmade process, salt brought through the irrigation water and accumulate in soil.

Secondary causes soil Salinity: In the secondary causes of soil salinization land is waterlogged by flooding irrigation water that affect the natural water balance of irrigated land. Most of water lost in this way and stored underground which can change the original water quality.

2. Problem caused by waterlogging conditions

There are three major problems available in saline soil for plant growth and development.

- **Oxygenation :** lack of oxygenation (due to filling of soil macro and micro pores) and decrease the crop production
- **Salt accumulated:** Salt accumulated through the salt affected irrigation water.
- **Seepage**: In this process, horizontal flow of water channel is called seepage and water loss through the irrigation channel.

3. Sources of Salinity

The major problem of soil salinity are accumulation of salt in soil (Through the irrigation of Salt affected water or mineral weathering) that create to problem for normal growth and development of plants. In this problem, Salts are a common and necessary component of soil, and many salts (e.g., Calcium and Magnesium) are essential plant nutrients. Salts originate from mineral weathering, inorganic fertilizers, soil amendments (gypsum, composts and manures), and irrigation waters.

4. Affected Area

Tube-wells and drains were installed in the most afflicted region, more than 3 million hectares of water-logged lands, at a cost of billions of rupees, although the reclamation objective was only partially achieved (Bhatti 1987). The Asian Development Bank (ADB) states that 38% of the irrigated area is now waterlogged and 14%of the surface is too saline for use (Bakker 2010). In the Nile delta of Egypt, drainage is being installed in millions of hectares to combat the waterlogging resulting from the introduction of massive perennial irrigation after completion of the High Dam at Assuan (Abdel-Dayem, 1987). FAO has estimated 106 ha of irrigated land will need to have improveddrainage systems installed, much of it subsurface drainage to control salinity (Biswas 1988).

5. Salinity Measurement

Soil salinity is measured by Electrical Conductivity (EC), solution obtained from water saturated soil paste. The unit of measuring salinity are deci Siemens per meter (dS/m).

6. Classification of salt affected soils

Table 1: Differentiate between saline, alkaline and saline-alkaline according to pH value, EC (dS/m) and ESP (%).

Parameter	Ph	EC (dS/m)	ESP (%)
Saline	< 8.5	> 4	< 15
Alkaline	> 8.5	< 4	> 15
Saline – Alkaline	> 8.5	> 4	> 15

Saline soil

The soils have an adequate amount of neutral soluble salts, which are harmful to the growth and development of the majority of agricultural crop plants. The soluble salts are mainly sodium chloride and sodium sulphate. However, significant amounts of Ca^{2+} and Mg^{2+} chlorides and sulphates are also present in saline soils. Electrical conductivity of the saturated extract (EC) more than 4 (dS/m) at 25 ℃ and exchangeable sodium percentage (ESP) less than 15% are characteristics of these soils (Table 1). These soils are typically in a flocculated form and are thought to have a permeability that is equivalent to or greater than that of regular soils due to the presence of excessive salts and low levels of Na^+ ions on exchange sites. The majority of these soils feature white encrustations of soluble salts or salt efflorescence near the surface.

Alkaline Soil

These soils contain sodium salts, primarily Na_2CO_3, that really can lead to alkaline hydrolysis. Older literature referred to these soils as "alkali" soils. At 25°C, the sodic soils exhibit EC less than 4 dS/m, ESP greater than 15, and SAR greater than 13 (Table 1). In these soils, majority of the sodium is available for the exchange. The soil solution contains very trace levels of free salts. pH of the soil is more than 8.5 these soils may become severely alkaline as a result of irrigation, and pH values that reach or exceed 10 are common. Sodic soils, often referred to as black alkali soils, have a brown-black appearance when organic matter is distributed and deposited on the surface.

Saline–alkaline

These soils are affected by sodium and other salts because they have a high concentration of neutral salts and a significant amount of sodium on the exchange complex. The Saline- alkaline soils have characteristics with EC above 4 dS/m at 25 ℃, ESP exceeding 15, and SAR more than 13. In these kinds of soils, both free salts and exchangeable Na^+ are found. Excessive salt content present in these soil keeps it flocculated, and its pH is often lower than 8.5. After leaching, due to the hydrolysis of exchangeable Na^+, these soils may act like a alkaline soil (pH > 8.5) when the free salt level decreases. Based on the expected salt damage to crops, the critical value for EC 4 dS/m, used to distinguish between a saline soil and a non-saline soil, was chosen. The yield of many crops is restricted at this point. Only sensitive crops see such a reduction in growth at EC levels between 2 and 4 dS/m. The effect of salinity is minimal below the EC value of 2 dS/m. Since no significant changes in soil characteristics have been noticed as the fraction of Na^+ ions on the exchange complex increases, the use of the ESP value of 15 is arbitrary. The ESP value of 15 has been utilized by the U.S. Salinity Laboratory as a boundary limit to differentiate sodic from non-sodic soils based on history and experience.

Chemistry between Salt-affected Soils

The series of the many events that leads to the formation of salt-affected soils begins with salt accumulation. Saline and alkaline soils develop in places with little rainfall (annual precipitation of less than 500 mm), when there is not enough precipitation to adequately drain away surface or ground water. Salts are progressively left behind in the soil, resulting to salt accumulation, while water is mostly lost in the atmosphere through evaporation or transpiration.

Salinization

Salinization is the process through which soluble salts accumulate in the soil. Among the salts, NaCl, Na_2SO_4, $CaCO_3$, and $MgCO_3$ predominate. Early salinization is frequently dominated by sodium salts. Salts of calcium and magnesium accumulate as time goes on. The soils produced are known as saline soils or white alkali soils; they were once also referred to as solonchaks. Soil salinity can also happen in soils that have been recovered from the seabed and soil in tide-affected coastal locations.

Alkalization or Sodication

When sodium-containing salts, especially its carbonates, are added to the soil, the soil's exchange complex may become saturated with sodium. Calcium and magnesium may precipitate as their respective carbonates when the salt concentration rises. It leads in a continuous rise in the amount of sodium in solution, which in turn causes an increase in the amount of sodium adsorbed on soil colloids and Calcium carbonate develops in the soils as a result. Sodication is the process of gradually raising the Na saturation on the soil exchange complex. Sodication causes physical properties to degrade, and the resulting soils are known as sodic soil, black alkali soils. Richards (1954) recommended the use of soluble salt and exchangeable sodium concentration as the two reliable criteria to identify saline and alkaline from other soils. The terminology (1) EC for soluble salts and (2) ESP for exchangeable sodium content are used to denote these characteristics. SAR (Sodium Absorption Ratio) is also employed to calculate soil sodicity.

Soil pH

In general, there is no clear correlation between the pH of the saturated soil paste and the soil ESP or SAR in saline soils. However, insodic soils or alkaline soil, there is a clear correlation between pH values and ESP of the soil or the SAR, so the pH can serve as a general indicator of the sodicity (Alkali) condition of the soil. However, a rise in soil pH is not always the consequence of soil sodification. Many sodic soils have a high concentration of neutral sodium salts like NaCl, making them neutral in response. Alkalization is caused by most sodic soil its strong alkaline reaction (pH=10). The latter is brought on by Na^+ ion or Na_2CO_3 molecule hydrolysis: The OH^- ions generated raise the pH of the soil because the exchange complex is saturated with Na^+ in soil. The latter may then go through hydrolysis, which also helps to raise the concentration of OH^- ions in the soil.

Electrical Conductivity (EC)

Electrical conductivity (EC) is measured to salt concentration in different soil varieties. Electrical conductivity differs occurs according to type of soil and pH also. The conventional measure of electrical conductivity of saline soil is dS/m. To eliminate the impact of temperature, it is common to test EC at a standard temperature of 25℃.

Exchangeable Sodium Percentage (ESP)

The percentage of sodium that has been adsorbed relative to the CEC is represented by the Exchangeable Sodium Percentage (ESP), which is written as follows:

$$ESP = E_{Na}X / \times 100$$

In terms of cation exchange capacity (CEC), Na^{2+}, Ca^{2+}, and Mg^{2+} represent the majority of the exchangeable cations in salt-affected soils.

Sodium Adsorption Ratio (SAR)

Most salt-affected soils have a high cation exchange capacity, with $Na2^+$, $Ca2^+$, and $Mg2^+$ making up the majority of the exchangeable cations.

$$SAR = \frac{Na^+}{\sqrt{(Ca^2 + +Mg^{2+})/2}}$$

Impact on Plant of Soil Salinity Stress

There are typical plant responses to various soil salinity stress ranges. Choose plants in these situations that can survive the soil's salinity level. Threshold values for harvested crops show soil salinity levels at which plants start to suffer any consequences that reduce output. Salinity values over the threshold that correspond to predicted yield losses of 10%, 25%, and 50% are occurred.

Table 2: Impact on plants under the salinity stress

S. N.	Salinity stress (dS/m)	Impact on plants
1.	0 – 2	No affect
2.	2 – 4	Growth of sensitive plants may be restricted
3.	4 – 8	Growth of many plants is restricted
4.	8 - 16	Only tolerants plants grow satisfactory
5.	>16	Very tolerant plants grow satisfactorily

9. Plant growth and developments under saline condition:

The need for irrigation for agricultural production in arid and semi arid regions has made the salt problem severe. At least 20% of all irrigated areas are estimated to be salt-affected (Pitman and Lauchli 2002). Irrigated agriculture accounts for more than 30% of all agricultural production, even though only around 17% of the area used for cultivation is irrigated **Hillel (2000).**Current estimate the entire worldwide area of salt-affected soils at about 830 million hectares (Martinez-Beltran 2005).

10. Salt sensitivity in relation to development plant growth stage

The sensitivity of a crop to salt changes from one developmental growth stage to the next has long been understood (Bernstein and Hayward (1958). The majority of studies shows that most annual crops are resistant at germination but sensitive during emergence and early vegetative growth, while there are few exceptions. Later in their growth and development, in particular, plants become more tolerant of saline as they grow older. These assertions are typically accurate (with the possible exception of a few crops), however it is crucial to stress that the concept of salt tolerance varies depending on the growth stage. While tolerance is often based on relative growth decreases throughout the later developmental phases, tolerance is typically based on % survival during germination and emergence. Depending on whether the harvested organ is a stem, leaf, root, shoot, fruit, fibre, or grain, salinity affects both vegetative and reproductive growth, which has significant implications. (Maas and Poss 1989).

11. Reclamation of Salt-affected Soils and its managements

Recovering the productivity of these soils requires reducing soluble salts and exchangeable sodium to levels that allow for optimal or nearly ideal plant development.

Physical Method of Saline Soil Reclamation

- **Scraping:** Mechanical methods might be employed to eliminate of the salts that have accumulated on the surface. If the area is extremely small, such as a small garden lawn or a patch in a field, this is the easiest and most cost-effective method of reclaiming saline soils.

- **Flushing:** washing away salts from the surface with freshwater. This is especially effective for crusted and impermeable soils. However, this is not a recommended plan of action.
- **Leaching:** The only effective method for removing excess salt from the soil is by leaching using fresh water, irrigation, or rain. If drainage facilities are available, it will work because draining the salt-rich wastewater will lower the water table and remove the salts.

Agronomical and Cultural Methods of Saline Soil Reclamation

The following cultural activities are used in this strategy in areas where there is only salty irrigation water available, where shallow saline water tables are widespread, and when soil permeability is poor.

12. Crop Selection and Rotations

The crops can be classified into four categories based on their tolerance to the salinity of the soil or the quality of irrigation water, such as

- **Highly tolerant crops:** Barley, Datepalm, Sugar beet etc.
- **Tolerant:** Coconut, Spinach, Amaranthus, Tapioca, Mustard, Pomegranate, Guava, and Ber etc.
- **Semi-tolerant:** Cabbage, Cluster bean, Pea, Lady's finger, Muskmelon, Onion, PotatoAsh guard, Bittterguard, Brinjal, Sweet potato, Tomato, Turnip, Water melon etc.
- **Sensitive:** Coriander, Cumin, Mint, Radish, Carrot, Grape, Sweet orange etc.

13. Raising Plants methods

- Rather of growing of crops from seeds, it is recommended to transplant seedlings (especially those of **vegetables, flowers, and fruit trees**).
- Grafted wild root stocks with a superior but salinity-sensitive scion **(Mango, citrus, Guava and ornamental plants like rose).**

14. Irrigation Practices

- Various method of water application- follow furrow or drip irrigation, and sprinkler irrigation and sub-surface irrigation systems.

- Irrigation frequency- irrigating more frequently (frequently) helps maintain greater water availability and reduce salinity, provided that there aren't too many irrigations.
- Use of mulching materials that prevent evaporation losses like straws, plastic sheets and other agricultural wastes etc.

15. Drainage Method of Saline Soil Reclamation: Drainage means removal of excess flooding water and along with it the salts from lands. In general drainage can be of two types viz.

16. Surface Drainage

Excess runoff in the form of flooding water is collected and drain out by the pipes and channels from the soil surface is called Surface drainage.

17. Subsurface Drainage

Gravitational water may be drained rapidly and efficiently using underground drainage canals. However, drainage ditches are difficult to use with machines and need thorough maintenance. Both of these drainage techniques have advantages and disadvantages depending on the unique local circumstances.

18 Chemical Methods

Gypsum application

Gypsum application was only necessary if the soils sodium concentration increased and the pH was more than 8 (as in saline sodic soil) so that Ca^{2+} could replace Na^+ and prevent Na^+ from subsequently draining out.

Nutrient Addition

Application of nutrients likes NPK, magnesium, and hormones like salicylic acid which increase optimum crop development and production while reducing the negative effects of saline soil. Potassium inhibits the absorption of Na while nitrate inhibits the uptake of chloride. Application of salicylic acid increases Mg intake, which affects ATP activity and increases H^+ATP-ase hydrolytic activity. It also increases H^+ ion import into the vacuole, which increases sodium absorption by the vacuole. By maintaining the water balance and ion ratio, K^+ foliar and soil treatment considerably minimizes the harmful effect of salty soil.

Biological Method of Saline Soil Reclamation

It is well known that as plants and cattle manures decompose in soil, carbon dioxide and organic acids are released. These gases aid in dissolving any insoluble calcium ions present in the soil solution as well as neutralize any alkali that may be present. Decomposing organic matter produces water-stable aggregates and improves soil permeability.

Organic Amendments for Improvement of Saline Soil

Farmyard manure, molasses, sugar industry press mud, green manures, crop residues, and various weeds are just a few examples of soil supplements that may be added to salinity-affected soil to help with the improvement and reclamation process over time. The amendment procedures are as follows.

Green Manuring and Crop Residues

There are some green manuring crops like- Dhaicha (*Sesbaniaaculeata*), Sunnhemps (*Crotalaria juncea*), Barseem (*Trifoliumalexandrieutn*), Sengi (*Melioletusparviflora*), and Cowpea (*Vignasinensis*), among others. They operate as sources of readily available nutrients during decomposition in addition to solubilizing calcium and lowering down the high pH of alkali soils. Increased biological activity and a resulting increase in soil permeability both contribute in the soil's sluggish regeneration. *Sesbaniaaculeata* has been discovered to be the most productive of all the plants utilized as green manure on salty soils. The following are crucial traits of crops that get green manuring: Alkalinity can be neutralized by legume crops. Highest percentage of calcium on an ash basis. Thrive in environments that are somewhat salinized.

Salty soil and water by Afforestation

Salt and water both are managed by afforestation. In this methods, Proper establishment of tree under favorable environment on salt affected soils and with brackish waters demands for the adoption of special package of practices. Before the afforestation, focused on some special points that are helpful for the afforestation programs.

- Salt Identification
- Quality of irrigation water and Assessment of availability
- Selection of tree species that are helpful
- Planting methods for salty soil
- Soil and water management
- Physical and social fencing during initial years

Conclusion

The majority of salt-affected soils are found in arid and semi-arid areas. Large-scale salt buildup in the root zone that has a negative impact on plant growth. However, many of these have the potential to be fertile soils and, if managed properly, can be quite productive. Based on the total salt concentration (EC) and the fraction of sodium among the actions, the USSL classification divides salt-affected soils into three types (EC, SAR and ESP). Saline and saline-alkaline soils have physical conditions that are satisfactory for plant growth, while alkaline soils have limited water permeability, are widely scattered, and are poorly aerated. The classification of salt-affected soils into saline and alkaline soils, which are separated by the predominance of chlorides and sulphates over sodium. The soil is saline if sodium is greater than chlorides and sulphates and the pH of the saturated soil paste is less than 8.2, and the soil is alkaline if sodium is less than chlorides and sulphates and the pH is greater than 8.2.By maintaining the root zone's salt and water balance by leaching. Making sure the soil has proper drainage is crucial. Gypsum ($CaSO_4.2H_2O$) must be applied as a suitable amendment to alkaline soil in order to remove sodium ions from the exchange complex before leaching may begin. Gypsum or any other amendment that contains calcium will promote flocculation and raise soil permeability, allowing sodium and other salts that have been replaced to be washed out from the root zone.

References

Abdel-Dayem, M. S., 1987. Development of land drainage in Egypt.In Proceedings, Symposium 25th Int.Course on Land Drainage. ILRI, publication 42.

Bakker, K., 2010. Privatizing water: governance failure and the world's urban water crisis. Cornell University Press.

Bernstein, L. and Hayward, H. E., 1958. Physiology of salt tolerance. Annual Review of Plant Physiology, 9(1):25-46.

Bhatti, A. K., 1987. A review of planning strategies of salinity control and reclamation projects in Pakistan. In Proceedings, Symposium 25th International Course on Land Drainage. ILRI, publication, 42.

Biswas, A. K., 1988. United nations water conference action plan: Implementation over the past decade. Int. J. of Water Resources Development, 4(3):148-159.

Hillel, D., 2000. Salinity management for sustainable irrigation: integrating science, environment, and economics. World Bank Publications.

Khatun, S. and Flowers, T. J., 1995. Effects of salinity on seed set in rice. Plant, Cell and Environment, 18(1):61-67.

Maas, E. V. and Poss, J. A., 1989. Salt sensitivity of wheat at various growth stages. Irrigation Sci., 10(1):29-40.

Martinez-Beltran, J., 2005. Overview of salinity problems in the world and FAO strategies to address the problem. In Managing saline soils and water: science, technology and social issues. Proceedings of the international salinity forum, Riverside, California, 311-313.

Pitman, M. G. and Lauchli, A., 2002. Global impact of salinity and agricultural ecosystems. In Salinity: environment-plants-molecules, 3-20.

13

Biofertilizer and their Importance for Sustainable Agriculture

Pramod Kumar[1], Ankit Tiwari[2] Akhilesh Nand Dubey[3] Navneet Maurya[4] and Himanshu Panday[5]

[1]Department of Soil Science and Agricultural Chemistry, SVPUAT Meerut, Uttar Pradesh
[2]Department of Agronomy, SVPUAT, Meerut, Uttar Pradesh
[3]Rabindra Ntah Tagore Agriculture College Deoghar, BAU, Ranchi
[4]Department of Agricultural Extension, CoA, BUAT, Banda, Uttar Pradesh
[5]YP– II, ICAR-Indian Institute of Sugarcane Research, Lucknow Uttar Pradesh

Abstract

The global expansion in human population poses a serious challenge to each person's food security, as agricultural land is finite and shrinking with time. As a result, it is necessary that agricultural production be significantly enhanced over the next few decades in order to supply the high food demand of the expanding population. Because of the growing population and decreasing agricultural area, there is a large imbalance between production and demand (7.2 million tonnes nutrient deficiency). To fulfil the expanding population's food need, agriculture production must be greatly increased during the next few decades. Chemical fertilisers have been widely used to close the gap between agricultural production and consumption, resulting in considerable harm to both nature and human health. Because of their considerable potential to increase food production and safety, biofertilizers are being used as an alternative for chemical fertilisers in the agriculture. Biofertilizers are chemicals that include cells of various beneficial microorganisms that have the potential to become essential components of enhanced nutrition management. Nitrogen fixers (N fixers), phosphorous solubilizers (P-solubilizers), and potassium solubilizers (K-solubilizers) are commonly used as biofertilizer ingredients, as is a mixture of fungi and moulds. The objective of this chapter is to explore the crucial functions of biofertilizers in sustainable farming.

Keywords: *Biofertilizers, Sustainable Agriculture, Food Security, Agricultural Resources etc.*

Introduction

Since the earliest stages of human evolution, agricultural and farming have been the only means of subsistence. The majority of people in the world depend on agriculture to support a healthy lifestyle and provide food, feed, and other essentials (such as fibre, wood, gums, and secondary products with medicinal properties) (Herve *et al.*, 2016; Kaur *et al.*, 2018). Given the anticipated increase in global population, industry and urbanisation are expanding quickly, causing environmental pollution that is becoming worse (Glick 2012). Additionally, the feeding the vast population at the moment which will definitely develop a big issue. However, the heavy use of chemical fertilisers in agriculture keeps the nation dependent on foreign food supplies while also seriously harming the environment and having a negative effect on all living things (Sujanya and Chandra 2011). By damaging the air, water, and soil, the unregulated use of chemical fertilisers poses a serious risk to the environment. Since the plants cannot absorb these dangerous compounds, they start to accumulate in ground water and a few of these substances are also to cause for the eutrophication of aquatic bodies. These substances have a negative impact on soil's ability to retain water, fertility, salinity, and distribution of nutrients (Savci 2012). With reference to the rising need for a continuous supply of nutritious food, long-term sustainability, and issues about environmental pollution, organic agriculture has emerged as an effective alternate sector by taking into consideration all the negative impacts of continuous use of chemical fertilisers. The success of sustainable agriculture is dependent on maintaining the environment, which is why sustainable agriculture principles go further than just growing crops to their maximum potential (Barragan-Ocana and Rivera 2016). As people have become more concerned with the preservation of natural resource and less interested with the farming systems as a whole, agricultural practises are developing in a strange way today. To increase crop productivity, agricultural operations utilise a variety of hormones, artificial fertilisers, and other synthetic minerals.

The health of the soil and the plant system are both impacted by artificial chemicals and minerals (Campos *et al.*, 2018). Although productivity may rise with increasing chemical usage, there are circumstances when loss of essential minerals and other nutritional variables behaves as a barrier with higher production. Without sacrificing the potential of future generations to fulfil their own needs and the environment's resources, sustainability in agriculture systems is possible (Wang *et al.*, 2015; Umesha *et al.*, 2018; Calabi-Floody *et al.*, 2018). Through acting as secondary pollutants, chemical fertiliser residues can penetrate food chains and food webs, eventually reaching humans, causing the depletion of favourable living circumstances. The impacts of secondary pollution on health hazards might cause them to remain in the environment for a comparatively longer time (Uosif *et al.* 2014). A new age of industry may begin if biofertilizers are used in place of pesticides in the agricultural system. Biofertilizers may be useful in giving agricultural plants the nutrients they need without causing environmental harm. The formulation of biofertilizers and their application to promote sustainability in agriculture may be helped by this chapter.

1. Biofertilizers

A biofertilizer is a material containing live microorganisms that, when given to seeds, plants, or soil, colonise the rhizosphere or within the plants and boost plant development by improving nutrient availability to the host plant. Biofertilizers are commonly used to enhance microbial activities that increase the supply of nutrients that plants can easily absorb. They increase soil fertility by biological nitrogen fixation and solubilizing insoluble phosphates, as well as synthesizing plant growth-promoting chemicals. Such biofertilizers have been pushed for their ability to extract the naturally occurring biological process of nutrient mobilisation, which greatly boosts soil fertility and, consequently, crop productivity. This chapter also discusses the potential of biofertilizers in many areas such as agriculture, ecology, and remediation, which may help to create biofertilizer as a useful strategy for sustainable agricultural development. Microorganisms are extracted from soil, water, air, or the rhizosphere for the creation of biofertilizers, which are then concentrated for use in the field. Microorganisms start developing agriculturally important metabolites in response to particular environmental circumstances, and these metabolites may be used by plants to support various biochemical processes (Salar *et al.*, 2017). Micro - organisms and microbiological metabolites assist the breakdown of complicated soil minerals into a simpler form that functions as a growth booster for a specific crop. Biofertilizers can be used for a wide range of uses.

2. Biofertilizers: Why is Their use Unavoidable?

Uncontrolled use of chemical fertilisers to feed the world's expanding population has surely contaminated the environment and seriously harmed beneficial insects' habitats. However, utilising too many chemical inputs has diminished soil fertility and increased the susceptibility of crops to disease. Agriculture must produce more food than there are accessible nutrients in order to feed the world's expanding population, and it must do it sustainably and environmentally friendly. The use of chemical fertilisers, pesticides, herbicides, fungicides, and insecticides is one of the many current agricultural practises that must be evaluated. Given the potentially harmful impacts of fertilisers, biofertilizers are meant to be a safe substitute for chemical inputs that significantly reduces ecological disturbance. Biofertilizers are affordable, environmentally benign, and when used over an extended period of time, they significantly improve soil fertility. According to reports, using biofertilizers increases crop output by 10 to 40 percent by boosting the number of proteins, vital amino acids, vitamins, and nitrogen fixation present in the soil. The advantages of adopting biofertilizers include affordable nutrient sources, superior micronutrient and microchemical suppliers, organic matter suppliers, the release of growth hormones, and the ability to reduce the negative effects of chemical fertilisers (Mahanty *et al.*, 2016).

3. Various Forms of Biofertilizers for Enhancing Crop Production

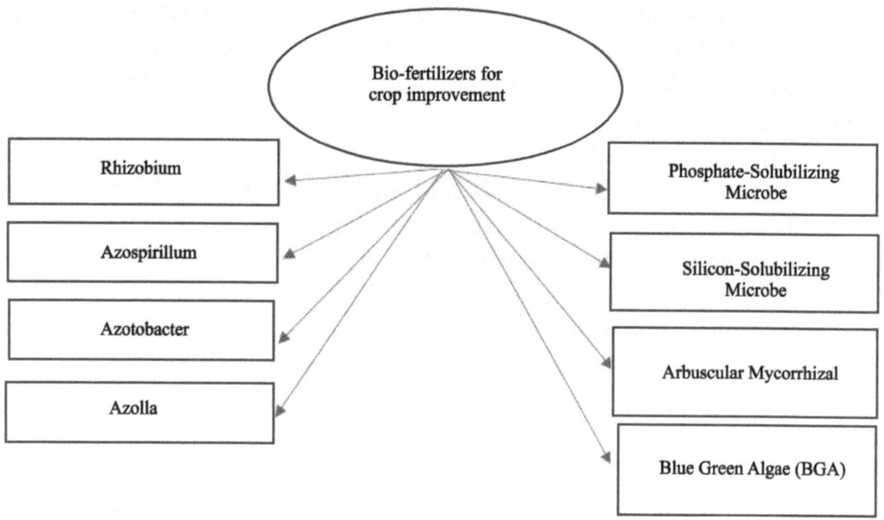

Fig. 1: Type of biofertilizers

4. Rhizobium

Symbiotic N_2-fixing rhizobacteria infect and form a symbiotic association with the root of leguminous. They are members of the *Rhizobiaceae* (-proteobacteria) family. This development involves a complex interaction between both the hosts and symbiont that results in the development of nodules and the colonisation of Rhizobia as an internal symbiont (Allito *et al.*, 2015). Rhizobia is the collective name for *Rhizobium, Bradyrhizobium, Sinorhizobium, Azorhizobium,* and *Mesorhizobium.* The term "diazotroph" refers to non-symbiont rhizobacteria that fixed nitrogen in non-leguminous plants and are also capable of developing a non-obligate relationship with their host plants (Verma *et al.*, 2010). The complex enzyme structure nitrogenase, which comprises of *dinitrogenase* reductase with iron (Fe) as a cofactor and *di-nitrogenase* with iron (Fe) and molybdenum (Mo) as its cofactor, is responsible for the nitrogen fixation process. The *nif* genes, which are present both in symbiotic and free-living nitrogen-fixing bacteria, are responsible for N_2-fixation (Black *et al.,* 2012). *Nif* genes include structural genes that are necessary for the synthesis and activity of the enzyme as well as regulatory genes that are involved in electron donation, Fe-Mo cofactor biosynthesis, and the activation of Fe proteins. Ultimately, when bean plants were infected with the modified Rhizobium, the plants displayed 68% higher nitrogenase activity than those that received the wild type *R. etli*. Due to this variation, the seeds produced had an increase in nitrogen of 16% and a 25–30% increase in leaf content. Nodule development is another significant Rhizobium-related factor. The majority of legume plants create "root nodules," de novo root lateral structures that host the symbiotic bacterium Rhizobium. This symbiotic bacterial infection of the legume plant changes the destiny of developed cortical cells throughout this phase, resulting in the creation of additional organs, such as nodules (Suzaki *et al.*, 2015).

To establish optimal nodule development, two regulatory processes, bacterial infection and nodule organogenesis, must be synchronized in the epidermal and cortical cells, respectively. The symbiotic reactions with host legume plants and Rhizobium were sustained by Nodulation factors (Nod factors), which are lipochitin oligosachcharides released by Rhizobia (Maillet *et al.*, 2011). It has been found that following Rhizobium sp. infection of legumes, the plant ethylene level increases, and this higher ethylene concentration can block further rhizobacterial infection and nodule development. Certain rhizobial strains can enhance the number of nodules formed on hosts legume plant roots by restricting the increase in ethylene production, which they achieve by generating a tiny chemical known as "rhizobitoxine" (Vijayan *et al.*, 2013). Rhizobitoxine is a phytotoxin that chemically prevents the ethylene biosynthesis enzyme 1-aminocyclopropane-1-carboxylate (ACC) synthase from carrying out its activity (Nascimento *et al.*, 2012). The enzyme ACC deaminase, which is produced by certain rhizobial strains, eliminates part of the ACC (the direct source to ethylene for plants) before it is transformed to ethylene. This decrease increases plant biomass by 25–40% and nodule development. Since only 1–10% of rhizobial strains in the field naturally contain ACC deaminase, genetically modifying these strain using Rhizobia ACC deaminase gene can improve the nodulation effectiveness of Rhizobia strains that do. *Azorhizobium*, one of several strains of Rhizobium, is a symbiotic bacterium that forms root nodes and fix nitrogen there. Additionally, they generate a lot of indole acetic acid (IAA), which stimulates plant development. *Bradyrhizobium* is a strong nitrogen fixer, and when the strain is used to inoculate mucuna seeds, it effectively increases the soil's total levels of organic carbon, N_2, phosphorus, and potassium. As a result, it greatly enhances plant development, the microbial population of the soil, and subsequently, plant biomass; it also decreases the population of weeds, etc (Mahanty *et al.*, 2016).

5. Azospirillum

They are gram-negative, aerobic nitrogen-fixing bacteria of the *Spirilaceae* family that don't form nodules (Mehnaz 2015). Although there are other species in this genus, including *Azospirillum amazonense, Azospirillum halopraeferens*, and *Azospirillum brasilense, Azospirillum lipoferum and A. brasilense* are the most helpful (Mishra *et al.*, 2013). While they cultivate that fix nitrogen on the organic salts of malic and aspartic acid, *Azospirillum* develops associative symbioses with many plants, especially those that have the C4 dicarboxylic pathway (Hatch-Slack pathway) of photosynthesis. As a result, it is mostly advised for the growth of crops including maize, sugarcane, sorghum, and pearl millet. They increase root growth and nutrient intake while also releasing growth promoters (IAA, gibberellins, and cytokinin) (N, P, and K). Root growth and exudation are significantly impacted by *azospirillum* inoculation. According to Steenhoudt and Vanderleyden, when *A. brasilense sp.* was introduced into maize, the organism's synthesis of several phytohormones rose noticeably, significantly enhancing maize growth. The maize plant's ability to absorb minerals was improved as a result of enhanced root physiology and architecture brought about by increased synthesis of several phytohormones. Nevertheless, Naiman *et al.*, (2009) demonstrated that the cultivable bacterial population in the wheat rhizosphere was altered by the inoculation of *Azospirillum* and *Pseudomonas*. Additionally, they noted that the profiles of the soil microflora's carbon source use during the tillering and grain filling phases were modified by the inoculation of *Azospirillum* and *Pseudomonas*.

6. Azotobacter

They are *Azotobacteriaceae*, which are free-living, aerobic, photoautotrophic, and non-symbiotic bacteria. The most prevalent species in arable soils is *Azotobacter chroococcum*, and they are often found in neutral and alkaline soils (Moraditochaee *et al.*, 2014). *Azotobacter vinelandii, Azotobacter beijerinckii, Azotobacter insignis,* and *Azotobacter macrocytogenes* are some of the other species that have been identified (Mishra *et al.*, 2013). Vitamin B complex, various phytohormones including gibberellins and naphthalene acetic acid (NAA), and other chemicals that inhibit some root infections while promoting root development and mineral absorption are secreted by them. According to reports, Azotobacter secretes certain chemicals that significantly enhance root development and nutrient intake while suppressing the growth of some root infections. The addition of 15–93 kg N/ha by Azotobacter to *Paspslum notatum* roots was also observed (Youssef and Eissa 2014). Another strain, *Azotobacter indicum*, is capable of producing a variety of antifungal antibiotics that are employed to reduce the growth of a number of harmful fungi in the root area and, in turn, somewhat lower seedling mortality. It has been discovered that the population of Azotobacter is often low in the soil and the rhizosphere of agricultural plants. Numerous agricultural plants, including rice, maize, sugarcane, bajra, vegetables, and plantation crops, have reported the presence of this bacterium in their rhizospheres (Wani *et al.*, 2013).

7. Blue-green Algae (BGA)

The nitrogen-fixing bacteria known as blue green algae (BGA) are filamentous in form and have specialised cells known as heterocysts (micronodules). Heterocysts exhibit nitrogen fixation process functioning. These microorganisms develop symbiotic partnerships with ferns, blooming plants, and fungal strains in order to fix nitrogen. Blue green algae play a significant role in agriculture because of their fast response and effective nitrogen fixation. In addition to fixing nitrogen, they also work to fix other micronutrients including phosphorus, zinc, potassium, sulphur, and others (Adeniyi *et al.*, 2018). Under the product names Klamath Blue Green Algae, Natural Blue Green Algae, Bulk Supplement Pure, and Blue Green Algae Pure Crystals, blue green algae are sold in powder form on the market.

8. Azolla

Azolla is a member of the *Salviniaceae* family, which includes seven diverse species of duckweed phototrophic fern (Roger and Ladha 1992). *Azolla* might develop a significant amount of biomass in just 10 days, depending on a number of variables, including the soil (pH, nutrients, type, and moisture). *Azolla* is a tiny super plant that floats freely and has distinctive scaly leaves and floating roots. The symbiotic nitrogen-fixing relationship between *Azolla* and *Anabaena azollae* is widely documented. Both developing and industrialised nations often employ azolla for nitrogen fixation (Emrooz *et al.*, 2018). The rice crop is noted for requiring more water than other crops, and farmers employ azolla to prevent excessive weed development. *Azolla*, although being free-floating, give developing rice plants up to 10 tonnes of protein and other necessary nutrients (Yao *et al.* 2018). Azolla is marketed as Hasiru green manure, urban farm, and biofertilizer among other names in the marketplace.

9. Phosphate-Solubilizing Microbe

Phosphorus is a macronutrient with particular significance since it controls plant respiration, protein synthesis, signal transduction, and nitrogen fixation (Khan *et al.*, 2010; Pande *et al.*, 2017). Because phosphorus is an insoluble component of soil, plants are unable to use it. It must be changed from its tied complex structure to its free form before frequent intake. To make phosphorus easier for plant roots to absorb, certain bacterial strains may transform to its most basic form. Phosphate-solubilizing bacteria, on the other hand, are common in nature; their abundance may differ based on the kinds of soil and the area from where they are collected. In developing countries, phosphate-solubilizing bacteria combined with cheap rock phosphate might replace expensive phosphate fertilisers (Mahanta *et al.*, 2018).

10. Silicon-Solubilizing Microbe

The soil profile may change as a result of natural processes including weathering of silica- and silica-based derivative-containing rocks (Kang *et al.* 2017; Vasanthi *et al.* 2018). A few different types of microbial consortia are crucial in the breakdown, transformation, and modification of silicon and its by-products. The capacity of microbial consortia to function depends on the soil's surrounding growth nutrients, pH levels, and availability of moisture. These are necessary for the synthesis of certain metabolites and enzymes that may aid in the mineral formation. The biological rather than the chemical transformation of hard silicon compounds into the most basic consumable forms has gained increased relevance. Biological approaches include the self-contained, affordable, and time-efficient actions of microbes that may change and modify an object.

11. Arbuscular Mycorrhizal

At various phases of growth and development, natural resources are constantly exposed to abiotic stressors. Plants begin synthesizing a specific category of secondary metabolites when stressed in order to battle the excess supply by reactive oxygen species (ROS) (Ahanger *et al.* 2014; Dhull *et al.* 2016). To some extent, the creation of certain ingredients supports the plant's survival under adverse conditions. One of the most essential variables that contribute to crop plant health is the symbiotic interaction. *Arbuscular mycorrhizal fungi* (AMF) are also an essential plant symbiont that aids in nutrient absorption and different enzymatic processes (Yang *et al.* 2018). AMF relationships with plant rhizospheres give a variety of growth-promoting effects such as better nutrition, increased resistance, drought tolerance, and modified soil composition (Gosling *et al.*, 2006; Berruti *et al.*, 2015). Organic farming avoids the application of water-soluble fertilisers and uses a number of crop rotations. According to scientific research, this results in increased AMF infection in soils with maximum nutrient absorption. As a result, AMF may be a viable alternative to chemical fertilisers.

12. Importance of Biofertilizer

Biofertilizers are required for restoring soil fertility. The use of chemical fertilisers over an extended period of time damages the soil and reduces crop production. Bio – fertilizers, on the other way, increase the soil's ability to hold water while also adding critical nutrients like nitrogen, vitamins, and proteins.

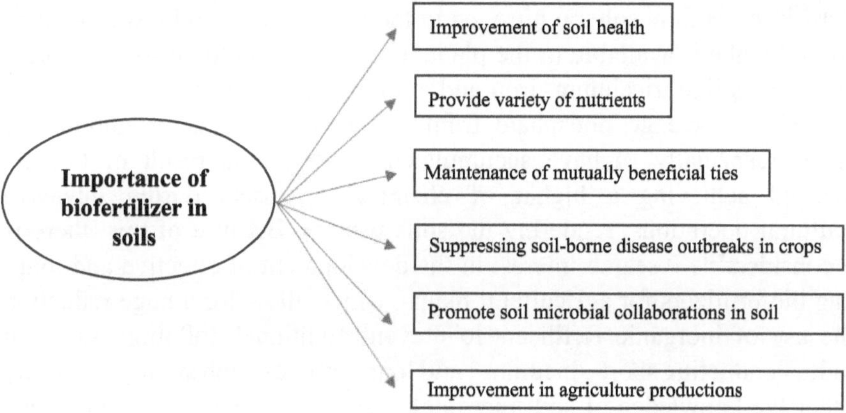

Fig. 2: Different type of improvement in soil

13. Characteristics required for release of biofertilizer

The adoption of biofertilizers by farmers to increase crop yield is one of the biggest challenges facing the agricultural industry. Even though there are several biofertilizers on the market right now, their amount and quality might differ based on the production facility. The following required characteristics should be present in a biofertilizer before it is released into the market.

Obtainability: Biofertilizers should be widely available. Farmers benefit from reduced transportation costs and saved time.

Storing constancy: Formulas for biofertilizers must be stable in a variety of atmospheric circumstances. With time, the formulation's quality needs to remain constant.

Efficiency: Biofertilizers should be successful in supplying a variety of nutrients that crops need while only requiring a little amount to be applied in the field.

Solubility and act: The composition should be water soluble to minimise overall costs and to allow for spray application in larger areas of the crop. The formulation should offer plants with a rapid supply of nutrients while having no negative effects. It should be simple to use and have no negative effects on the farmer's health. It should be affordable to farmers since it has an impact on agricultural prices. It should be season-independent and available to growers the entire year.

Conclusion

Biofertilizers fix atmospheric nitrogen in the soil and root nodules of legume crops and make it available to the plant. They solubilise the insoluble forms of phosphates like tricalcium, iron and aluminium phosphates into available forms. They scavenge phosphate from soil layers. Excessive amounts of nutrients (especially P) have accumulated in soils as a result of farmers unchecked achieving a higher of chemical fertilisers during intensive agricultural operations, rendering the soils useless. Because of this, there is now considerable research interest in the development of effective and long-lasting biofertilizers for agricultural plants, which allow for a huge reduction in the use of inorganic fertilisers to prevent additional pollution issues. It includes conducting short-, medium-, and long-term research studies that bring together the knowledge of soil microbiologists, agronomists, plant breeders, plant pathologists, nutritionists, and economics.

References

Adeniyi, O. M., Azimov, U. and Burluka, A., 2018. Algae biofuel: current status and future applications. Renew Sust Energ Rev, 90:316–335.

Ahanger, M. A., Hashem, A., Abd-Allah, E. F. and Ahmad, P., 2014. Arbuscular mycorrhiza in crop improvement under environmental stress. In: Ahmad P, Rasool P (eds) Emerging technologies and management of crop stress tolerance, Elsevier, New York, 2:69–95.

Allito, B. B., Nana, E. M. and Alemneh A. A., 2015. Rhizobia strain and legume genome interaction effects on nitrogen fixation and yield of grain legume: a review. Molecular Soil Biology, 6:1–6.

Barragan-Ocana, A. and Rivera, M. C., 2016. Rural development and environmental protection through the use of biofertilizer's in agriculture: an alternative for underdeveloped countries. Technol Soc., 46:90–99.

Berruti, A., Lumini, E., Balestrini, R. and Bianciotto, V., 2015. Arbuscular mycorrhizal Fungi as natural biofertilizers: let's benefit from past successes. Front Microbiol, 6:1–13.

Black, M., Moolhuijzen, P., Chapman, B., Barrero, R., Howieson, J., Hungria, M. and Bellgard, M., 2012. The genetics of symbiotic nitrogen fixation: comparative genomics of 14 Rhizobia strains by resolution of protein clusters. Genes 3(1):138–166.

Calabi-Floody, M., Medina, J., Rumpel, C., Condron, L. M., Hernandez, M., Dumont, M. and Mora, M. L., 2018. Smart fertilizers as a strategy for sustainable agriculture. Adv Agron,147:119–157.

Campos, E. V. R., Proenca, P. L. F., Oliveira, J. L., Bakshi, M., Abhilash, P. C. and Fraceto, L. F., 2018. Use of botanical insecticides for sustainable agriculture: future perspectives. Ecol Indic, 4:29-38.

Dhull, S. B., Kaur, P. and Purewal, S. S., 2016. Phytochemical analysis, phenolic compounds, condensed tannin content and antioxidant potential in Marwa (Origanum majorana) seed extracts. Resour Effic Technol, 2:168–174.

Emrooz, H. B. M., Maleki, M. and Rahmani, A. (2018) Azolla-derived hierarchical nanoporous carbons: from environmental concerns to industrial opportunities. J Taiwan Inst Chem Eng, 91:281–290.

Glick, B. R., 2012. Plant growth promoting bacteria: mechanisms and applications. Scientifica, 6:15-23.

Gosling, P., Hodge, A., Goodlass, G. and Bending, G. D., 2006. Arbuscular mycorrhizal fungi and organic farming. Agric Ecosyst Environ, 113:17–35.

Herve, M., Albert, C. H. and Bondeau, A., 2016. On the importance of taking into account agricultural practices when defining conservation priorities for regional planning. J Nat Conserv, 33:76–84.

Kang, S., Waqas, M., Shahzad, R., You, Y., Asaf, S., Khan, M. A., Lee, K., Joo, G., Kim, S. and Lee, I., 2017. Isolation and characterization of a novel silicate-solubilizing bacterial strain Burkholderia eburnea CS4-2 that promotes growth of japonica rice (Oryza sativa L. cv. Dongjin). Soil Sci Plant Nutr, 63(3):233–241.

Kaur, P., Dhull, S. B., Sandhu, K. S., Salar, R. K. and Purewal, S. S., 2018., Tulsi (Ocimum tenuiflorum) seeds: in vitro DNA damage protection, bioactive compounds and antioxidant potential. J Food Meas Charact,12:1530–1538.

Khan, M. S., Zaidi, A., Ahemad, M., Oves, M. and Wani, P. A., 2010. Plant growth promotion by phosphate solubilizing fungi current perspective. Arch Agron Soil Sci., 56:73–98.

Mahanta, D., Rai, R. K., Dhar, S., Varghese, E., Raja, A. and Purakayastha, T. J., 2018. Modification of root properties with phosphate solubilizing bacteria and arbuscular mycorrhiza to reduce rock phosphate application in soybean-wheat cropping system. Ecol Eng, 111:31–43.

Mahanty, T., Bhattacharjee, S., Goswami, M., Bhattacharyya, P., Das, B., Ghosh, A. and Tribedi, P., 2016. Biofertilizers: a potential approach for sustainable agriculture development. Environ Sci Pollut Res, DOI 10.1007/s11356-016-8104-0.

Maillet, F., Poinsot, V., André, O., Puech-Pagès, V., Haouy, A., Gueunier, M., Cromer, L., Giraudet, D., Formey, D., Niebel, A. and Martinez, E. A., 2011. Fungal lipochitooligosaccharide symbiotic signals in Arbuscular mycorrhiza. Nature 469(7328):58–63.

Mehnaz, S., 2015. Azospirillum: a biofertilizer for every crop. In Plant microbes' symbiosis: applied facets, Springer India, 297–314.

Mishra, D. J., Singh, R., Mishra, U. K., Kumar, S. S., 2013. Role of bio-fertilizer in organic agriculture: a review. Res J Recent Sci., 2:39–41.

Moraditochaee, M., Azarpour, E. and Bozorgi, H. R., 2014. Study effects of biofertilizers, nitrogen fertilizer and farmyard manure on yield and physiochemical properties of soil in lentil farming. International Journal of Biosciences 4(4):41–48.

Naiman, A. D., Latrónico, A. and de Salamone, I. E., 2009. Inoculation of wheat with Azospirillum brasilense and Pseudomonas luorescens: impact on the production and culturable rhizosphere microlora. Eur J Soil Biol, 45(1):44–51.

Nascimento, F. X., Brígido, C., Glick, B. R. and Oliveira, S., 2012. ACC deaminase genes are conserved among Mesorhizobium species able to nodulate the same host plant. FEMS Microbiol Lett, 336(1):26–37.

Pande, A., Pandey, P., Mehra, S., Singh, M. and Kaushik, S., 2017. Phenotypic and genotypic characterization of phosphate solubilizing bacteria and their efficiency on the growth of maize. J. Genet Eng. Biotechnol, 15:379–391.

Roger, P. A. and Ladha, J. K. (1992) Biological N-2 fixation in wetland rice fields: estimation and contribution to nitrogen balance. Plant Soil, 141:41–55

Salar, R. K., Purewal, S. S. and Sandhu, K. S., 2017. Bioactive profile, free-radical scavenging potential, DNA damage protection activity, and mycochemicals in Aspergillus awamori (MTCC 548) extracts: a novel report on filamentous fungi. Biotech, 7:164-178.

Savci, S., 2012. An agricultural pollutant: chemical fertilizer. Int. J. of Envi. Sci. and Dev., 3(1):73-82.

Sujanya, S. and Chandra, S., 2011. Effect of part replacement of chemical fertilizers with organic and bio-organic agents in ground nut, Arachis hypogea. J. of Algal Biomass Utilization 2(4):38–41.

Suzaki, T., Yoro, E. and Kawaguchi, M., 2015. Chapter three-leguminous plants: inventors of root nodules to accommodate symbiotic bacteria. Int. Rev. Cell Mol Biol, 316:111–158.

Umesha, S., Manukumar, H. M. G. and Chandrasekhar, B., 2018. Sustainable agriculture and food security. In: Biotechnology for sustainable agriculture emerging approaches and strategies. Woodhead Publishing, Cambridge, MA, 67–92.

Uosif, M. A. M., Mostafa, A. M. A., Elsaman, R. and Moustafa, E. S., 2014. Natural radioactivity levels and radiological hazards indices of chemical fertilizers commonly used in upper Egypt. J. Radiat Res. Appl. Sci., 7:430–437.

Vasanthi, N., Saleena, L. M. and Raj, S. A., 2018. Silica Solubilization potential of certain bacterial species in the presence of different silicate minerals. SILICON, 10:267–275.

Verma, J. P., Yadav, J., Tiwari, K. N. and Lavakush, S. V., 2010. Impact of plant growth promoting rhizobacteria on crop production. Int. J. Agric. Res., 5:954–983.

Vijayan, R., Palaniappan, P., Tongmin, S. A., Padmanaban, E. and Natesan, M., 2013. Rhizobitoxine enhances nodulation by inhibiting ethylene synthesis of Bradyrhizobium elkanii from lespedeza species: validation by homology modeling and molecular docking study. World J.of Pharmacy and Pharmaceutical Sci., 2:4079–4094.

Wang, H., Liu, S., Zhai, L., Zhang, J., Ren, T., Fan, B. and Liu, H., 2015. Preparation and utilization of phosphate biofertilizer's using agricultural waste. J. Int. Agric. 14:158–167.

Wani, S. A., Chand, S. and Ali, T., 2013. Potential use of Azotobacter chroococcum in crop production: an overview. Current Agric. Res., 1(1):35–38.

Yang, H., Schroeder-Moreno, M., Giri, B. and Hu, S., 2018. Arbuscular mycorrhizal Fungi and their responses to nutrient enrichment. In: Giri B, Prasad R, Varma A (eds) Root biology. Soil biology. Springer, 52:429–449.

Yao, Y., Zhang, M., Tian, Y., Zhao, M., Zeng, K., Zhang, B., Zhao, M. and Yin, B., 2018. Azolla biofertilizer for improving low nitrogen use efficiency in an intensive rice cropping system. Field Crop Res., 216:158–164.

Youssef, M. M. and Eissa, M. F., 2014. Biofertilizers and their role in management of plant parasitic nematodes. A review. J. of Biot. and Pharmaceutical Res., 5(1):1–6.

14

Recommended Doses of Fertilizers in Horticultural Crops

***Km Kusum*[1]*, Ajay Hemdan*[2]*, Sampurna Nand Singh*[3]*, Kuldeep*[4] *and Purnita Raturi*[5]**

[1,3,4&5]*Department of Horticulture, CoA, GBPUAT, Pantnagar, Uttarakhand*
[2]*Department of Medicinal and Aromatic Plants, CoH, VCSGUUHF Bharsar, Uttarakhand*

Abstract

The recommended dose of fertilizer in horticultural crops is a critical factor influencing crop yield and quality. Various factors, including soil type, crop species and climate, contribute to the complexity of determining the ideal fertilizer dosage. The recommended dose of fertilizer is not a one-size-fits-all solution, emphasizing the need for soil testing to assess nutrient levels. Problems arise when excessive or insufficient fertilizers are applied, leading to nutrient imbalances, environmental pollution and economic losses. To address these challenges, precision agriculture technologies, such as satellite imaging and sensor-based systems, can be employed to optimize fertilizer application based on real-time crop and soil conditions. The effects of improper fertilization on horticultural crops include reduced yields, poor fruit quality and increased susceptibility to diseases. Moreover, environmental concerns, such as nutrient runoff and groundwater contamination, highlight the importance of adopting sustainable fertilization practices. Strategic approaches involve incorporating organic matter, utilizing cover crops and employing controlled-release fertilizers to enhance. Nutrient availability while minimizing environmental impact. A tailored approach to fertilizer application in horticultural crops is crucial for sustainable agriculture. By integrating soil testing, precision agriculture technologies and environmentally conscious practices, farmers can optimize crop nutrition, mitigate adverse effects, and promote long-term soil health. A balanced and informed fertilization strategy is essential for ensuring the resilience and productivity of horticultural crops while, minimizing environmental consequences. Ongoing research and collaboration between researchers, farmers and policy makers are imperative to refine and promote best practices in horticultural crop fertilization.

Keywords: *Fertilization, Nutrition, Precision agriculture, Policy, Soil Testing etc.*

Introduction

The horticultural industry includes fruits (such as almonds), vegetables (such as potatoes, tuber crops, and mushrooms), and ornamental plants (such as cut flowers and spices). With plantation crops and aromatic and medicinal plants accounting for 30.4% of the nation's agricultural GDP, they have become a major force behind economic growth. Potato, tomato, onion, brinjal, and cabbage account for 80% of production in vegetables, while banana, mango, and citrus make up 70% of fruit crops. The following states: Tamil Nadu, Bihar, Gujarat, Karnataka, Madhya Pradesh, Uttar Pradesh, Maharashtra, West Bengal. A significant portion of horticulture produce comes from Odisha. Fertilizer application in horticultural crops has enormous potential. Admittedly, the application of fertilizer to horticultural crops has frequently been random and arbitrary. Mango and guava are two examples of fruit crops that seldom ever receive fertilizer. However, a study conducted on vineyards in Maharashtra and Karnataka reveals that overuse of fertilizers, totaling to 3 t (3000 kg) ha-1, has resulted in ground water contamination, the development of saline in the poorly drained soils, and a loss of fruit quality (Chadha, 1989). Fertilizer, whether of natural or synthetic origin and excluding liming materials, is a substance applied to soil or plant tissues, typically leaves, to provide essential plant nutrients for growth. Essentially, fertilizers are materials with a specific chemical composition, possessing higher analytical value, and capable of supplying plant nutrients in readily available forms. These industrially manufactured chemicals contain a higher concentration of nutrients compared to organic manures, and the nutrients are quickly released upon application (Havlin *et al.,* 2016). The recommended dose of fertilizers refers to the quantity or amount of fertilizers that should be applied to a specific area of land or to a particular crop to meet the nutritional requirements of the plants. It is an essential aspect of modern agricultural practices to ensure optimal plant growth, development, and yield. The recommended dose takes into account factors such as soil type, crop type, climate conditions, and specific nutrient needs. Application of fertilizers to plantation, fruits and vegetable crops, though not uncommon, has not been addressed adequately. Based on the recommended doses of fertilizers, Ghosh, 1999 projected the N, P_2O_5 and K_2O requirement for the year 2000 at 826, 570 and 115 thousand tonnes for different fruit crops, respectively. Similarly, Singh and Kalloo, 1999 projected these requirements for vegetable crops at 744,000 tonnes for N, 372,000 tonnes for P_2O_5 and 272,000 tonnes for K_2O. Thus, fruit and vegetable crops together had a potential requirement of 2.90 Mt of major nutrients in 2000 AD.

Beginning of 21st century has witnessed a significant jump in the area, production and productivity of horticultural crops. Particularly the area increased from 15 Mha in 2001-02 to 25.7 Mha in 2017-18 and the production witnessed a quantum jump from 146 Mt to 306.8 Mt. Proportionately the fertilizer consumption has also increased and there is a need to reassess the requirement of fertilizers for horticulture sector. Horticultural crops have a tremendous potential for fertilizer use. Nutrient uptake by many fruits and vegetable crops are comparable with those of cereal crops to replenish the removal and to supply sufficient amount of nutrients at each stage of crop growth, adequate rates are needed in the fertilizer application programmed of horticultural crops. Different horticultural crops and have presented the growth trends by putting up the data for 2003-04 and 2017-18. In the year of 2003-04 the requirement of nitrogen fertilizer for is 7,38,857, P_2O_5: 3,93,081 and 6,83,162 tonnes K_2O was recorded for fruit crops. Similarly in the year of 2017-2018 it was increased *i.e.,* 16,23,217, 10,15,340 and 17,10,995 tonnes. In similar manner the N: P: K fertilizer requirement increased year by year. In 2003-2004 the N: P: K fertilizer requirement for vegetable crops 4,75,215, 4,50,648 and 5,06,230 was recorded. It increase double or three times increase in the year of 2017-2018. These estimates show that today horticulture sector requires about 9.0 Mt of NPK fertilizers. The decadal trends have shown that the consumption of NPK fertilizers in horticulture sector increased by 112%. Individual sector-wise, the fruit crops showed an increase of 140%, vegetables of 93%, and plantation crops of 18%. Though the growth rates appear to be impressive, many orchards and fields under horticultural crops continue to remain neglected (Ganeshmurthy *et al.,* 2019).

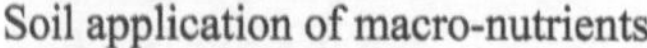

Soil application of macro-nutrients

Foliar application of micronutrients

Foliar application of Micronutrients

- **Classification of fertilizers**

(A) Based on the number of nutrients supplied by a particular fertilizer

1. **Straight fertilizers:** Straight fertilizers are those which supply only one

primary plant nutrient, name nitrogen or phosphorus or potassium *eg.* urea, ammonium sulphate, potassium chloride and potassium sulphate.

2. **Complete fertilizers:** Complete fertilizers is referred to fertilizer material which contain all three major nutrients eg. N:P:K.

3. **Incomplete fertilizer:** This fertilizer is referred to a fertilizer material which lacks anyone of three major nutrient elements.

4. **Complex fertilizers/ Compound fertilizers or mixed fertilizers:** Complex fertilizers contain two or more primary plant nutrients of which two primary nutrients are in chemical combination. These fertilizers are usually produced in granular form *e.g.* Di ammonium phosphate nitro phosphates and ammonium phosphate

(B) Fertilizers can also be Classified Based on Physical Form

1. Solid fertilizers: Solid fertilizers are agricultural fertilizers in a solid, granular, or powdered form. These fertilizers are typically applied to the soil to provide essential nutrients to plants, promoting their growth and enhancing crop yields. Solid fertilizers come in various formulations and compositions, each designed to address specific nutrient requirements of different crops or soil conditions. Here are some common types of solid fertilizers:

A. Nitrogen Fertilizers

- **Urea:** A popular nitrogen fertilizer with high nitrogen content. It is suitable for a wide range of crops.
- **Ammonium sulfate:** Contains nitrogen and sulfur, suitable for crops that require both nutrients.

B. Phosphorus Fertilizers:

- **Superphosphate:** Contains phosphorus in a form that is readily available to plants. It is often used for promoting root development.
- **Triple superphosphate**: Contains a higher concentration of phosphorus compared to superphosphate.

C. Potassic Fertilizers:

- **Potash (potassium chloride):** Provides potassium, an essential nutrient for overall plant health and disease resistance.
- **Sulfate of potash:** Contains both potassium and sulfur.

Solid fertilizers are applied to the soil either by broadcasting over the field or by placing them in proximity to the plant roots during planting. The choice of fertilizer depends on factors such as the soil type, crop requirements, and the specific nutrient deficiencies present. Additionally, proper application rates and timing are crucial for maximizing the effectiveness of solid fertilizers and minimizing environmental impact.

2. Liquid fertilizers: Liquid fertilizers are formulations of essential nutrients that are dissolved or suspended in a liquid carrier. These fertilizers are commonly used in agriculture, horticulture, and gardening to provide plants with the necessary nutrients for growth. Liquid fertilizers offer certain advantages over solid fertilizers, such as quicker nutrient absorption and the ability to provide a more precise application eg. Anhydrous ammonia.

Liquid fertilizers can be applied through various methods, including foliar spraying, soil drenching, and fertigation. Foliar application allows for rapid nutrient uptake through the leaves, while fertigation involves applying the fertilizer through irrigation systems. The choice of liquid fertilizer depends on the specific needs of the plants, the growing conditions, and the desired method of application. Proper dilution rates and application timing are important considerations to ensure effective and efficient nutrient delivery to the plants.

There are 25.7 Mha of horticultural crops on the land. 15.3 Mha of this area is covered with plantations and perennial horticulture crops, such as fruits, nuts, and spices. Annual Horticultural crops are very separate from annuals in that their growth stages are totally different and their rates of nutrient absorption correspond to their growth patterns. Table 1 presents general fertilizer requirements for various categories of horticulture crops. Based on this data, the N: P_2O_5: K_2O ratio of suggested fertilizer A variety of horticultural crops were calculated. It's surprising how much the advice differs from one another compared to other commercial and agricultural crops. Spices receive higher potassium (K) in relation to nitrogen (N), followed by plantation crops, spices, fruit crops, and vegetable crops, which receive the least amount of K. These ratios are very different from those of commercial and field crops, and the published data don't provide strong evidence for why. The researchers should pay attention to this area of concern.

The Recommended Dose of Fertilizer of some Horticultural Crops is given below

Table 1: The recommended dose of fertilizer of fruit crops

S.No	Crops	RDF NPK (Kg ha^{-1})		
		N	P_2O_5	K_2O
1.	Banana	620	310	620
2.	Mango	75	20	70

S.No	Crops	RDF NPK (Kg ha^{-1})		
		N	P_2O_5	K_2O
3.	Citrus	110	35	55
4.	Papaya	925	925	925
5.	Guava	250	175	175
6.	Apple	320	320	320
7.	Pineapple	275	70	200
8.	Sapota	100	50	50
9.	Grapes	300	300	600
10.	Pomegranate	500	425	975
11	Litchi	50	50	25

Table 2: The recommended dose of fertilizer of vegetable crops

S.No	Crops	RDF NPK (Kg ha^{-1})		
		N	P_2O_5	K_2O
1.	Potato	60	100	120
2.	Tomato	180	120	150
3.	Onion	125	75	125
4.	Brinjal	180	150	120
5.	Tapioca	45	90	120
6.	Cabbage	150	125	100
7.	Cauliflower	150	100	100
8.	Okra	100	50	50
9.	Peas	25	75	60
10.	Sweet potato	20	40	60
11	Chilli	150	75	75

Table 3: The recommended dose of fertilizer of spices

S.No	Crops	RDF NPK (Kg ha^{-1})		
		N	P_2O_5	K_2O
1.	Garlic	40	75	75
2.	Turmeric	150	60	108
3.	Ginger	37.5	50	37.5
4.	Cumin	30	20	20
5.	Coriander	10	40	20
6.	Tamarind	20	15	25
7.	Fenugreek	30	25	40
8.	Fennel	50	10	10
9.	Pepper	110	50	155
10.	Cardamom	75	75	150
11	Ajwain	40	20	20
12	Nutmeg	187.5	187.5	600

Table 4: The recommended dose of fertilizer of plantation crops

S.No	Crops	RDF NPK (Kg ha^{-1})		
		N	P_2O_5	K_2O
1.	Coconut	100	55	210
2.	Cashew nut	100	40	60
3.	Areca nut	140	55	200
4.	Cocoa	70	30	100

Sources: IIHR (2012); Muvel, et al., (2015)

Conclusion

Fruit, vegetables and plantation crops need much more fertilizer than agricultural crops, so the horticulture industry has a lot of opportunity to grow. Growers have more challenges to how much fertilizer should be applied to their horticulture crops. Extremely productive cultivars may produce large yields. Using suitable crop protection and fertilizer management techniques on productive soils with sufficient water availability. Horticultural crops that are annuals constantly assimilate nutrients; therefore it is beneficial to supply them with properly balanced levels of nutrients throughout their growth. Applying slow-release fertilizers, which are made expressly to release nutrients over a lengthy period of time, is the most effective approach to accomplish this. However, there is additionally apprehension that the disproportionately widespread application of fertilizers in horticulture and their low efficiency. This raises the risk of environmental degradation, lowers crop resistance to pests and diseases, and decreases the quality of horticultural produce. Today's horticulture farmers and proprietors of businesses must adopt balanced fertilization practices determined by comprehensive soil and plant analyses.

References

Chadha, K.L., 1989. Keynote address on potassium is nutrition of fruits and vegetable crops in India. In Group Discussion on Potassium and its Influence on Quality of Fruit and Vegetable Crops. Held on 2nd December, at the Indian Agricultural Research Institute, New Delhi.

Ganeshamurthy, A. N., Kalaivanan, D., Rupa, T. R., & Manjunath, B. L., 2019. An assessment of the fertilizer needs of horticultural crops in India. Indian Journal of Fertilisers, 15(3): 286-295.

Ghosh, S. P., 2000. Nutrient management in fruit crops. Fertiliser News, 45(4): 71-76.

Havlin, J. L., Tisdale, S. L., Nelson, W. L., & Beaton, J. D., 2016. Soil fertility and fertilizers. Pearson Education India.

IIHR. 2012. Production Technology of Tropical Fruit Crops: A Hand Book. IIHR, Bengaluru.

Muvel, R., Naruka, I. S., Chundawat, R. S., Shaktawat, R. P. S., Rathore, S. S., & Verma, K. S., 2015. Production, productivity and quality of ajwain (Trachyspermum ammi L. Sprague) as affected by plant geometry and fertilizer levels. International Journal of Seed Spices, 5(2): 32-37.

Singh, K. P., & Kalloo, G., 2000. Nutrient management in vegetable crops. Fertiliser News, 45(4): 77-81.

15

Analysis of Cost and Returns Efficiency: Applying Sen's MOP Model for Potato Cultivation of Nalanda District in Bihar

***Suman Kumari*[1]*, Mukesh Maurya*[2]*, Akash Rai*[3] *and Priyanka Kumari*[4]**

[1&2]*Department of Agricultural Economics, SHUATS, Naini, Prayagraj Uttar Pradesh*
[3]*Agri-Business Management, SHUATS, Naini, Prayagraj, Uttar Pradesh*
[4]*Department of Agricultural Economics, SHUATS, Naini Prayagraj Uttar Pradesh*

Abstract

India is the world's second–larger producer of veggies, after China. The primary vegetable grown in Nalanda is the potato. The goal of the current analysis was to analysis potato farming costs and returns using various cost models. The study was carried out in Bihar sarif and Noor Sharai block, which were chosen due to their high potato production and area. Thus, the probability proportion to the number of farmers in each size and categories were taken for the analysis for the cost of cultivation of potato of the sample farms in Nalanda district of Bihar. Production is normally considered as the function of area and yield. The selection of the crop enterprise to be chosen on the farm and the distribution of space and resources under it depends heavily on the crop's yield, price, and the cost of the inputs used to produce it. These measurements for potato were developed as a result of the cultivation costs and returns to various production aspects being taken into consideration when choosing a crop. Per hectare, on an average Rs. 92552.03 was spent on potato. The most money was spent on farming in the medium farm category Rs. 100166, followed by small farms Rs. 92867 and marginal farms Rs. 84623.09. The cost of cultivation's various components in variable cost like tuber (seed) accounted for the highest amount of (22.95%), followed by (19.46 per cent) Human labour. The remaining significant factor included the fertilizer (8.82 per cent), irrigation charge 7.77 per cent, machine charge 6.49 per cent, plant protection 3.79 per cent, manure 3.06 per cent, depreciation rate 1.62 per cent and land revenue 0.24 per cent. And cost of fixed cost in highest amount of percentage rental value of own land (21.61 per cent), interest on fixed capital 2.95per cent and interest on working capital 1.23 per cent. The risks involved in agriculture urges the need for insulating the farm profits through income optimization management. The study has attempted an evolving suitable strategy for increasing farm income and employment through vegetable crops with less use of irrigation water in eastern Uttar Pradesh. The objective was fulfilled using Sen's Multi-Objective Programming (MOP) model. The estimates of MOP have been found as a better solution over the individual optimization but as a compromised one. The optimal cropping plan was also found superior over the existing cropping plan for increasing income and employment with lesser irrigation. An appropriate resource use planning is required for the development of farming sector. Increasing income, employment with lesser use of irrigation, fertilizer etc. may be the major objectives to be achieved. There are several mathematical models for achieving multiple objectives. Sen's Multi Objective Programming (MOP) model is most popular for achieving several objectives simultaneously. In the present study, an optimal cropping plan was proposed for Potato growers of Nalanda district of Bihar for increasing income, employment with lesser use of fertilizer.

Keywords: *Cost of Cultivation, Production, Net Returns, Farm Business Maximization, Linear Programming, Multi- Objective Programming etc.*

Introduction

The vast Indo- Gangetic plains, one of the world's most productive regions, are represented by Bihar. It provides ideal agro- climatic conditions for raising a variety of crops. Approximately 80% of the workforce in Bihar is employed in agriculture, which also accounts for nearly 42% of the state's GDP. Due to the fact that 89.5% of the state's population has a rural or farming background, Bihar contributes 39% of the nation's agricultural GDP as opposed to the average of 24.3% for the entire country. With a per capita income of roughly one – fourth of the country, Bihar ranks third in people and tenth in area. With 83 million people and an annual population growth of roughly 2.43 %, Bihar is the most populous state in India. After rice, wheat, and maize, potatoes are the fourth most important food crop in Bihar. With output ranking just fourth after rice, wheat and maize, it takes up less than 5 % of the net seeded area. High in nutrients, simple to digest, and recognized as a healthful crop with enormous potential for assuring food and nutritional security for millions of people significant issues in Bihar. This crop was determined to be the most labour- and capital-intensive because of the high costs of seed, fertiliser, and human labour. Human labour alone made up over 35% of the Rs. 135317 total costs, backed by seed (23%). The return input ratio over the total shelled out cost was therefore 1:1.39. The investigation revealed that by optimising the use of labour, manures, and fertilisers, there is significant potential to increase the profit from the potato harvest (Sharma et. al., 2017).

The objective of an efficient farming system is to develop optimal combinations of enterprises for greater integrations of farming activities. Agricultural planning has to be carefully organized at the farm level itself to make better use of the agricultural land of the region, to sustain farming families and to produce crops on a sustainable basis. Most of the farmers of the area follow diversified farming system. They grow multiple crops in a season to meet the consumption needs of their families as well as generate sufficient income for other household expenses. Therefore, enhancement of farmer's income was made an important consideration for formulating alternative farming plan. Farmers are using less organic manure in their crops and more chemical fertilizers, which is harmful for the soil health. The reduction of the use of chemical fertilizers was another consideration for appropriate farm planning. More opportunities for employment for the rural laborers should also be generated through improved farm planning.

The application of multi objective optimization is required to achieve multiple objectives simultaneously. Sen's MOP model (Sen 1994) is very popular and efficient in providing an optimal solution with multiple conflicting objectives. This method has been successfully used in many research studies (Gangwar 1994 and Gautam 2013), (Kumar 2012 and Kushwaha 1992), (Kumari 2017), (Maurya 2019) and (Sen 1983) for formulating the suitable farm plans to achieve several objectives simultaneously.

Results and Discussion

The cost incurred by various size groups in various potatoes growing activities is shown down in table 1 by size group. Potato costs per hectare were, on average, Rs 92552.03. small farms (Rs 92867) and marginal farms (Rs 84633.09) had the lowest cultivation costs, while medium farms had the highest, at Rs 100166. The cost of cultivation's various components in variable cost like tuber (seed) accounted for the highest amount of (22.95%), followed by Human labor (19.46 per cent). The remaining significant factor included the fertilizer (8.82 per cent), irrigation charge 7.77 per cent, machine charge 6.49 per cent, plant protection 3.79 per cent, manure 3.06 per cent and additional cost in needed during potato farming.

Total cost of cultivation (Cost C2) per hectare of onion amounted to Rs. '72258,'Rs. 83689 and Rs. '91535 on small, medium and large farm, respectively with an average of Rs. '77850. On an average, Cost A1 was Rs. 52519, which was highest on large farms (Rs. '76937) and low on small farms ('40963). The average Cost B1 and B2 were Rs. '53525 and Rs. '63525 respectively. Among different land size categories, Cost C1 was highest (Rs. '81595) on large farms and lowest (Rs. '62259) on small farmers with an average of Rs. '67850. Cost C3, which includes managerial cost, was worked out to be Rs. '85635 per hectare on an overall basis. An increasing trend was observed in different costs with increases in the farms size. (Susheela Meena et. al., 2016).

Table 1: Cost of input of potato cultivation in different size of farms holdings

Items	**Marginal**	**%**	**Small**	**%**	**Medium**	**%**	**Sample Average**	**%**
Variable cost								
Family Laboure	12270	14.50	11874	12.79	3314	3.31	9152.67	9.89
Hired	4576	5.41	7099	7.64	14900	14.88	8858.33	9.57
Machinecharge	4753	5.62	5775	6.22	7502	7.49	6010.00	6.49
(Tubers)Seed	16552	19.56	21155	22.78	26012	25.97	21239.67	22.95
Manure	2405	2.84	2740	2.95	3349	3.34	2831.33	3.06
Fertilizer	8114	9.59	8066	8.69	8312	8.30	8164.00	8.82
Plant protection measures	3168	3.74	3535	3.81	3827	3.82	3510.00	3.79
Irrigation charger	7751	9.16	7211	7.76	6624	6.61	7195.33	7.77
Depreciation	1347.37	1.59	1548	1.67	1602	1.60	1499.12	1.62
Land revenue	230.52	0.27	225	0.24	210	0.21	221.84	0.24
Total variable Cost	48896.89	57.78	57354	61.76	72338	72.22	59529.63	64.32
Fixed cost								
Interest on fixed capital@ 11%	2271.73	2.68	2481	2.67	3442	3.44	2731.58	2.95
Rental value of own land	20000	23.63	20000	21.54	20000	19.97	20000.00	21.61
Interest on workingcapital @7.5%	1184.47	1.40	1158	1.25	1072	1.07	1138.16	1.23
Total Fixed Cost	23456.2	27.72	23639	25.45	24514	24.47	23869.73	25.79
Total Cost	84623.09	100	92867	100	100166	100	92552.03	100

All costs were comparatively higher for large farm followed by medium, small and marginal farmers. It shows that capital spending on production increased with increase in the farm size. This was because the large farmers purchased qualitative inputs in each and every season which were required for production of potato and also income source at large farms enabled them to purchase costlier inputs and hiring the Laboure for performing different activities in potato cultivation. (Sinha. Aswin et.al., 2019).

Table 2: Cost and returns in potato production

Particulars	Marginal	Small	Medium	Sample Average
Cost of Cultivation (Rs. /ha)	84623.09	92867	100166	92552.03
Production (Q/ha)	191.31	196.67	202.92	196.97
Price (Rs/ha)	1115.06	1220.1	1300	1211.72
Gross returns (Rs. /Ha)	213322.13	239957.07	263796	238668.45
Net returns (Rs. /ha)	128699.04	147090.07	163630	146473.04
Cost of production (Rs. /Q)	442.33	472.20	493.62	469.39

Table 2 indicates that the potato production cost and return with growing farm size of households, it was observed that parameters like cost of cultivation, gross return, and cost of production of potatoes were rising. However, marginal, small, and medium potato growers, respectively, had net returns on production of Rs 128699.04, Rs 147090.07, and Rs 163630 of potatoes. This showed that medium farmers used their resources the best, followed by small farmers and finally marginal potato growers. The yield of potatoes per hectare was, on average, 196.97 quintals. The figure shows that, on average, potato growers made Rs. 238668.45 per hectare in net income from potato farming.

Table 3: Cost of cultivation of potato on different cost concepts basis on different size holdings (Rs. /ha.)

Particulars	Marginal (<1 ha)	Small (1-2 ha)	Medium (<2-10 ha)	Sample Average
Cost A1	51168.62	59835	75780	62261.21
Cost A2	51168.62	59835	75780	62261.21
Cost B1	52353.09	60993	76852	63399.36
Cost B2	72353.09	80993	96852	83399.36
Cost C1	64623.09	72867	80166	72552.03
Cost C2	84623.09	92867	100166	92552.03
Cost C3	93085.40	102153.7	110182.6	101807.23

The data in table 3 indicates that on marginal, small and medium farms, respectively, the total cost of cultivation (C2) for potatoes per hectare was Rs 84623.09, Rs 92867, and Rs 100166, cost A1 was, on average, Rs 62261.21; The figure shows that the average cost to produce one hectare of potatoes in the study region was calculated at Rs. 62261.21, of which cost Al, which is a sum of all variable costs, came to be estimated at Rs 62261.21 per hectare of the overall expenses (Cost C3). Cost Al increased to Rs. 63399.36 per hectare after interest on fixed capital was added. This expense is referred to as cost B1 of the total expense. Cost B2 of the total cost was Rs. 83399.36. Cost C1, which is the sum of cost B1 and the imputed value of family labour, was determined to be Rs. 72552.03. Cost C2, which made up Rs. 92552.03 per acre of the total cost, is made up of Cost B2 and the imputed value of family labour. The total cost, or cost C3, which is comprised of cost C2 and 10% of cost C2 as managerial costs, was calculated to be Rs. 101807.23 per acre total, which includes administrative costs. As the size of the farm increased, an upward tendency in various costs was seen.

Income Measures

A comparison of various income measures from potato cultivation in Nalanda District in Bihar Potato offers positive benefits to farmers of all farm size groups as far as farm business income, family labour income and farm investment income, return over variable cost, return per Rupees and Return to management are concerned.

Table 4: Returns from cultivation of potato crop on sample farms (Rs./ha.)

Particulars	Marginal	Small	Medium	Sample Average
Return over variable Cost	162153.51	180122.07	188016	176763.86
Farm business income	162153.51	180122.07	188016	176763.86
Family labour income	140969.04	158964.07	166944	155625.70
Farm investment income	149883.51	168248.07	184702	167611.19
Return per rupees	2.52	2.58	2.63	2.58
Return to management	120236.73	137803.37	153613.4	137217.83

As shown in table 4 the range of return over variable cost was from Rs. 162153.51 to Rs. 188016. As the size of the land holding expanded, the returns over variable costs also increased. there was no difference between cost A1 and A2 because land leasing for vegetable production was not common in the research area, therefore farm business revenue, which represents returns over costs A2, was the same as returns over variable cost. From Rs 140969.04 on marginal farms to Rs 166944 on medium farms, the family labour income per hectare of potato cultivation varied and the total household labour income was calculated to be Rs 155625.70 per hectare. The total farm investment income on an average Rs 167611.19 per hectare. On a cost-3 basis, the total returns to management from potato farming were Rs. 137217.83 per hectare. On various land size holding, it ranges from Rs 120236.73 to Rs 153613.4.

Net returns on Different Cost Concept Basis

Table 5: Net returns per hectare from potato cultivation on different cost concept (Rs./ha.)

Particulars	Marginal (<1 ha)	Small (1-2 ha)	Medium (2-10 ha)	Sample Average
Cost A1	162153.51	180122.07	188016	176763.86
Cost A2	162153.51	180122.07	188016	176763.86
Cost B1	160969.04	178964.07	186944	175625.70
Cost B2	140969.04	158964.07	166944	155625.70
Cost C1	148699.04	167090.07	183630	166473.04
Cost C2	128699.04	147090.07	163630	146473.04
Cost C3	120236.73	137803.37	153613.4	137217.83

It is clear from above table that the comparative estimates of different costs incurred in potato cultivation for different size group are giving in table 5. Potato offers positive benefits to farmers of all farm size groups as far as farm business income, family labour income and farm investment income, return over variable cost, return per Rupees and Return to management are concerned, as shown in Table 5, the range of return above variable cost was from Rs. 162153.51 to Rs 188016. As the size of the land holding expanded, the returns over variable costs also increased. there was no difference between cost A1 and A2 because land leasing for vegetable production was not common in the research area, therefore farm business revenue, which represents returns over costs A2, was the same as returns over variable cost. From Rs 140969.04 on marginal farms to Rs 166944 on medium farms, the family labour income per hectare of potato cultivation varied. The total household labour income was calculated to be Rs 155625.70 per hectare B2. On a cost C3 basis, the total returns to management from potato farming were Rs. 137217.83 per hectare. On various land size of holding, it ranges from Rs 120236.73 to Rs 153613.4. all together, the returns from costs A1, A2, B1, B2, C1, C2 and C3 were, respectively, Rs 176736.86, Rs 176763.86, Rs 175625.70, Rs 155625.70, Rs 166473.04, Rs 146473.04 and Rs 137217.83 per hectare of potato production.

Table 6: Return per rupees of investment in potato cultivation

Particulars	Marginal (<1 ha)	Small (1-2 ha)	Medium (2-10 ha)	Sample Average
Cost A1	4.17	4.01	3.48	3.89
Cost A2	4.17	4.01	3.48	3.89
Cost B1	4.07	3.93	3.43	3.81
Cost B2	2.95	2.96	3.29	3.07
Cost C1	3.30	3.29	3.29	3.29
Cost C2	2.52	2.58	2.63	2.58
Cost C3	2.29	2.35	2.39	2.34

One of the best ways to gauge the economic viability of any crop is return on investment in rupees. For the cultivation of potato as shown in table 6 clears out that the average values for the C1 C2 and C3 were 3.89, 3.89, 3.81, 3.07, 3.29 and 2.34, respectively. Medium farms yielded the best returns per rupees of investment on a cost – per – unit (C2) basis (2.63), followed by marginal (2.58) and small (2.52). The results demonstrated that medium farms were more effective than small – sized and marginal sized farms, mostly due to lower cost per unit of output.

Table 7: Cost of production of potato different size holding (Q/ha.)

Particulars	Marginal (<1 ha)	Small (1-2 ha)	Medium (2-10 ha)	Sample Average
Cost A1	267.46	304.24	373.45	315.05
Cost A2	267.46	304.24	373.45	315.05
Cost B1	273.66	310.13	378.73	320.84
Cost B2	378.20	411.82	477.29	422.44
Cost C1	337.79	370.50	395.06	367.79
Cost C2	442.33	472.20	493.62	469.39
Cost C3	486.57	519.42	542.99	516.32

In table 7 shows how much it costs to produce potato across various categories of land size. According to costs C2 basis, it is revealed that a quintal of potatoes cost, on average, 469.39 to produce. Production costs were 315.05, 315.05, 320.84, 422.44, 367.79 and 516.32. A1, A2 B1, B2, C1 and C3 basis, respectively and C2 is main cost of production 2.58. is increasing trend marginal, small and medium group.

To Assess the Optimum Utilization of Available Resources in Potato Cultivation

The household in question owns 1.51 hectares of land used for the cultivation of paddy, maize, red gram, potato, wheat, and lentils. The estimated net income per hectare for the household was as follows: 62938/ha for rice, 37753/ha for maize, 31724/ha for red gram, 90055/ha for potato, 53127/ha for wheat and 35613/ha for lentils. The household is looking for crop combinations that was assist them optimize their annual net profits overall. Whether this crop enterprise producing mix is ideal is of utmost importance. Does it produce the highest net returns? Land, labour, and operating capital are the resources that were taken into consideration in this research model.

X1	=	Area allocation for rice crop
X2	=	Area allocation for maize crop
X3	=	Area allocation for red gram crop
X4	=	Area allocation for potato crop
X5	=	Area allocation for wheat crop
X6	=	Area allocation for lentil crop
X7	=	labour hiring
X8	=	Capital borrowing

Table 8: MOP matrix

Crop	**Income (ha)**	**Land (ha)**	**Labour Manday (ha)**	**Operating Capital (ha)**	**Fertilizer (kg/ha)**	**Top 20% (Maximum area/ha)**	**Bottom 20% (minimum area/ha)**
Kharif							
Rice (x1)	62938	0.61	58	35000	230	1.85	0.16
Maize (x2)	37753	0.44	41	20000	190	1.31	0.09
Red gram (x3)	31724	0.45	43	15000	85	0.97	0.18
Rabi							
Potato(x4)	90055	0.84	63	55000	240	2.22	0.18
Wheat(x5)	53127	0.40	56	30000	180	1.09	0.10
Lentil(x6)	35613	0.27	37	12000	60	0.71	0.07

Resources used in MOP model

a. Kharif Land (1.51/ha)

b. Rabi Land (1.51/ha)

c. Kharif Labour (120-man days/ha)

d. Rabi Labour (120-man days/ha)

e. Wage Rate (Rs.350-man days)

f. Rate of Interest on borrowing 7% per annum

g. Kharif own capital 48,000

h. Rabi own capital 48,000

Formulation Linear Programming (LP) Model

Maximize Z = 62938 X1 + 37753 X2 + 31724 X3 + 90055 X4 + 53127 X5 + 35613 X6 – 350 X7 – 350 X8 – 0.07 X9 - 0.07 X10

Subject to

Crop Land Constraints = $X1 + X2 + X3 \leq 1.51$ (Kharif Season)

= $X4 + X5 + X6 \leq 1.51$ (Rabi Season)

Labour constraints = $58\ X1 + 41\ X2 + 43\ X3 \leq 120$ (Kharif Season)

= $63\ X4 + 56\ X5 + 37\ X6 \leq 120$ (Rabi Season)

Operating Capital Constraints

$35000\ X1 + 20000\ X2 + 15000\ X3 \leq 48,000$ (Kharif Season)

$55000\ X1 + 30000\ X5 + 12000\ x6 \leq 48,000$ (Rabi Season)

Applying the simplex procedure for obtaining the optimum land of Food Crops through Temporally-Ordered Routing Algorithm (TORA) computer-based software.

All the three objective functions have been achieved using linear programming. The results of individual optimization are presented in Table 9. It is clear from the table that maximization of income has increased in income by 12.98 percent over its existing level. However, the employment increased by 4.09 percent and fertilizer use increased by 7.96 percent. The employment maximization has increased by 7.02 percent over its existing level.

Table 9: Individual and Multi-objective optimization

Item	Existing Level	Max. of Income	Max. of Employment	Min. of FertilizerUse	Multi-Objective Optimization
Income (Rs)	184592	208549 (12.98)	176491 (-4.39)	143138 (-22.46)	204556 (10.82)
Employment (Days)	171	178 (4.09)	183 (7.02)	146 (-14.62)	181 (5.85)
Fertilizer Use (Kg.)	616	665 (7.96)	604 (-1.95)	543 (-21.86)	634 (2.92)

Note: Figures in parentheses shows the percentage increase/decrease over the existing levels.

However, the income and employment has decreased by 4.39 and 1,95 percent respectively over their existing levels. The minimization of fertilizer use decreased the fertilizer use by 21.86 percent with decrease in income and employment by 22.46 and 14.62 respectively over their existing levels. None of the above mentioned three solutions are favourable and acceptable to the farmers of Nalanda district. However, Sen's MOP has generated a compromise solution as mentioned in last column of the table. The income level of farmers has increased by 10.82 percent over its existing level and very close to its individual maxima of Rs. 208549. The employment has also increased by 5.85 percent over its existing level. However, the fertilizer use has not decreased but increased by 2.92 percent only. The minor increase in fertilizer use is acceptable with favourable increase in income and employment in the proposed cropping plan. The existing and optimal cropping plan is mentioned in table 10. The farmers in the study area were growing six major crops of Paddy, Maize and Red Gram in Kharif season and Potato, Wheat and Lentil in the Rabi season in their average holding of 1.51 hectares.

Table 10: Existing and Optimal Cropping Pattern

Crops	Existing Area (ha.)	Optimal Area (ha.)
Paddy	0.61 (20.26)	0.30 (9.93)
Maize	0.44 (14.62)	0.99 (32.78)
Red Gram	0.45 (14.95)	0.22 (7.29)
Potato	0.84 (27.91)	1.18 (39.07)
Wheat	0.40 (13.29)	0.20 (6.62)
Lentil	0.27 (8.97)	0.13 (4.31)
Sown by the Gross Cropped Area	3.01 (100.00)	3.02 (100.00)

Paddy in kharif and Potato in rabi season were the major crops grown by the farmers of the study area. However, in the proposed cropping plan, the area under Paddy and Red Gram has decreased significantly and shifted in the area of Maize. There is significant increase in the area of Maize from 0.44 ha. in the existing cropping plan to 0.99 ha in the proposed cropping plan. Similarly, in the Rabi season the area under Potato has increased from 0.84 ha. in the existing cropping plan to 1.18 ha. in the proposed cropping plan. The area under Wheat and Lentil has reduced significantly in the proposed cropping plan.

The application of multi objective optimization is required to achieve multiple objectives simultaneously. Sen's MOP model (Sen 1994) is very popular and efficient in providing an optimal solution with multiple conflicting objectives. This method has been

Successfully used in many research studies (Maurya et. al., 2019) for formulating the suitable farm plans to achieve several objectives simultaneously.

Conclusion

Potato cultivation costs tended to rise as holding size increased. On medium farms, as compared to small and marginal farms, the yield per hectare was higher. As a result, medium farms had greater gross returns per hectare of potato farming. The cost of cultivation analysis reveals that for the sample farms in the study area, the average total (cost C2) per hectare of potato production was RS. 92552.03 medium farms had the greatest C2 costs, followed by small and marginal farms. The total gross income from potato farming per hectare was Rs. 238668.45. In comparison to small and marginal farms, this was higher on medium – sizes farms. The pattern was the same for the family's labour – based income. The household labour income per hectare, on an average, was Rs. 155625.70. potato production generated that medium farm made the most money per hectare, followed by small and marginal sized farms. The average cost of production on the sample farms was (Rs 469.39). on an average, 2.58 rupees were earned foe every rupee invested. It was highest on (2.63) medium farms, (2.58) small farms and (2.52) marginal farms respectively.

The present study indicated the possibilities of improving farming practices for the welfare of the farmers. The existing farming plan can be altered for achieving multiple objectives using Sen's MOP technique. The proposed cropping plan improved the income and employment as desired. However, the fertilizer used has not reduced as desired. The improvement in fertilizer application techniques is required.

References

https://cacp.dacnet.nic.in

Kumari, M., Singh, O.P. and Meena, D.C., 2017. Optimizing cropping pattern in Eastern Uttar Pradesh using Sen's Multi-Objective Programming Approach. AERA, 30(2):

Maurya, M., Kamalvanshi, V., Kushwaha, S. and Sen, C., (2019) Optimization of Resources Use on Irrigated and Rain-fed Farms of Eastern Uttar Pradesh: Sen's Multi-Objective Programming (MOP) Method. Int. J. Agric. Stat. Sci. 15(1): 183-186.

Sen, C.,1983. A new approach for multi-objective rural development planning. The Ind. Eco. J. 30(4): 91-96.

Shankar, T., Singh, K.M., Kumar, A. and Singh, S.K., 2014. Cultivation and Processing of Potato in Bihar: Issues and Strategies. Environment & Ecology.32(4B): 1647—1652.

Sharma, V., Lal, H., Debnath, U. and Hatte, V., 2017. Economics of Potato Production in Kangra District of Himachal Pradesh, India. Int. J. of Current Micro. and Appl. Sci. 6(10): 123-12.

Singh, B., Prasad, V., Gupta, B. K., and Kumar, B., 1998. Economic effects of storage in marketing of potato in district Farrukhabad (UP) Bihar J. of Argic. Marketing 6(4) 179-184

Sultana, M.S., Kabir, F., Islam, M.S., Rashid, M.M. and Akon, A.I., 2005. Economic study on winter vegetables produced by different categories of farms. J. of Biological Sci. 5(2): 107-110.

16

Millets
Holistic Approach Towards Health

Rekha Singh[1] and Rahul Kumar Rai[2]***

[1]*Krishi Vigyan Kendra, Bhadohi, Uttar Pradesh*
[2]*Department of Agricultural Economics, CoA, BUAT, Banda, Uttar Pradesh*

Abstract

The increasing an importance of millets about health, nutrition, food security, inconvenience in food preparation and lack of processing industry. In India, millets are generally consists of pearl millet, finger millet, kodo millet, foxtail millet, sorghum, barnyard, proso and little millet. These are used as traditional food for centuries. Millets contain important nutrients such as antioxidants like phenolic acid, flavonoids, phytosterols, lignins and lipids etc. In developing countries, millets are mainly grown for fodder purposes. These millets are also considered as gluten free cereals. Thus, it is best suited for people who are gluten intolerant. Sorghum is fifth main crop after wheat, paddy, maize and barley. It is mainly grown in central India such as Maharashtra, Andhra Pradesh and Telangana. Pearl millet is widely grown in Rajasthan and Gujarat and Finger millet is the main cultivated coarse grain in Karnataka, Gujarat and Tamil Nadu. The consumption of millets reduces many health risks like heart diseases, poor digestive system, prevent diabetes, neural problems, low energy levels with the help of different meal. Apart from this, millets have potential from protection against age related degenerative diseases. In barnyard millet has sufficient amount protein which is highly digestible and good for cardiovascular diseases and diabetes patients. Barnyard millet is found efficient in reducing blood glucose and lipid levels. Lecithin presents in kodo millet is useful for strengthening nervous system. Kodo millet is good source of vitamin B group and minerals like calcium, potassium, iron, zinc and magnesium. It is very beneficial for postmenopausal women who are suffering from cardiovascular diseases, high blood pressure and high cholesterol levels. Pearl millet is good source of iron so it is very beneficial for pregnant women. So, it can be say that consumption of millets protects from many health and age related degenerative diseases. In rural areas, many traditional methods of processing millets are used to prepare food items according to local taste. These traditional methods will have to be made suitable for larger market keeping in mind the taste of others. Market can be created by upgrading old processing techniques with new processing techniques. So that food items made from millets can be available to everyone.

Keywords: *Millets, Nutrition, Food Security etc.*

Introduction

Millets are first plants domesticated for food and generally refer to cereals that are considered nutritious and offer various health benefits. These cereals are often rich in essential nutrients like vitamins, minerals, fibre and sometimes protein. Including a variety of nutri-cereals in your diet can contribute to overall well-being and support a balanced nutritional intake. Millets are well-regarded for their nutritional content, climate resilience, and suitability for sustainable agriculture. Millets, also referred as nutri- cereals are a group of small-seeded grasses that have been cultivated for thousands of years and are an essential part of the staple diet in many parts of the world, especially in Asia and Africa. They are gaining popularity globally due to their nutritional benefits, environmental sustainability, and gluten-free nature. In recent years, there has been a growing awareness of the nutritional benefits of millets, leading to increased promotion and support for millet cultivation by government initiatives and programs. These initiatives aim to improve the livelihoods of farmers, enhance food security, and promote environmentally sustainable agricultural practices. There are several types of millets like pearl millet, sorghum, finger millet, proso millet, kodo and barnyard millet, each with its own unique nutritional profile. In India, millets are cultivated from ancient times especially in eastern Uttar Pradesh. Pearl millet or spiked millet is the main cultivated crop in eastern Utter Pradesh among other millets. Sorghum is also cultivated as fodder crop. Finger millet was cultivated in some regions but due to its tedious dehusking process, farmers stopped its cultivation. Lower rate of consuming millets at local level is one of the main causes of malnutrition.

As of now, well informed with nutrition benefits of millets, urban people have started including millets in their diet regularly. Different types of flour and food items made with millets are also available in the market in different packaging. But in the villages, farmers are not much aware about nutrition aspects of millets and this is the main reason that farmers are not including millets in their diet at regular basis. In winters, they used only pearl millet in the form of flour. To increase consumption of millets at local level, there must be some training and technologies like ready to cook, ready to eat, instant food, millets based local snacks can be provided to farmers. Policy makers can focus on skill development programs based on millets for farm women to increase in their income and well being.

In eastern Uttar Pradesh, flaking, popping of pearl millets and sorghum has been in practices. Rural people have used this as snacks. With changing eating habits, extruded products are in demand but the main problem is availability of millet based products. Extruded products are made with refined wheat flour is available but these are not healthy and consumption at large level spoils people's health.

Some of the Most Common Types of Millets

1. **Pearl Millet (Pennisetum glaucum)**: Also known as Bajra in India, Pearl Millet is a widely grown millet variety. It is a good source of protein, iron, fibre and essential nutrients. It is commonly used in various dishes, including flatbreads and porridge.
2. **Finger Millet (Eleusine coracana)**: Popularly known as Ragi in India, Finger Millet is rich in calcium, iron and fibre. It is often used to make flour for flatbreads, porridge and other traditional dishes.
3. **Sorghum (Sorghum bicolour)**: Sorghum also known as Jowar is a versatile millet with a high nutritional value. It is rich in antioxidants, vitamins and minerals. Sorghum is used to make flour for flatbreads, porridge and various other food products.
4. **Foxtail Millet (Setaria italica)**: Foxtail Millet is an ancient grain with a mild flavour. It is rich in dietary fibre, protein and essential minerals. It is commonly used in the preparation of rice substitutes, dosa and various other dishes.
5. **Proso Millet (Panicum Miliaceum)**: Proso Millet is known for its quick growth and adaptability to different climatic conditions. It is a good source of carbohydrates, protein and fibre. Proso Millet can be used in porridge salads and as a rice substitute.
6. **Little Millet (Panicum Sumatrens)**: Little Millet is small-grained millet rich in fibre, iron and other nutrients. It is commonly used in South Indian cuisine to make idlis, dosas and porridge.
7. **Kodo Millet (Paspalum Scrobiculatum)**: Kodo Millet is rich in protein, fibre and antioxidants. It is used to make a variety of dishes including porridge, upma and rotis. Kodo Millet is also known for its ability to withstand dry and arid conditions.

8. **Barnyard Millet (Echinocholoa esculenta)**: Barnyard Millet is fast growing millet with high nutritional value. It is rich in fibre, protein and essential nutrients. It can be used in various dishes such as porridge, pulao and dosas.

Nutritional Profile of Millets

Millets are rich in phytochemicals such as polyphenols, phytosterols, phytocynanins, and lignans. These function as immune boosters, antioxidants and detoxifying agents. These attributes protect against age related degenerative diseases like diabetes, cancer and cardiovascular diseases. Millets are also a rich source of vitamin B group. In terms of nutrition, millets surpass rice and wheat offering high amount of calcium, protein and fibres. Millets are non acid forming so they encourage alkalinity in the body. Including an alkaline based diet is advised for gaining optimal health. As we know that finger millet is an excellent source of calcium. Finger millet can be included as supplementary food for children and daily intake of this can reduce chances of osteoporosis in women. Pearl millet is rich iron and protein. Weakness in the body, fatigue, energy deficiency all these are a common health issues that farm women generally told about. If farm women include pearl millet or a combination of millets in their diet daily, that can be a great help in removing all these symptoms. Finger Millet is especially beneficial for its potential to address malnutrition and provide essential nutrients. In Karnataka state, ragi slurry is given to infants for better growth and development.

Sorghum consumption for long time protects from many health diseases such as obesity, diabetes, celiac disease and prevents from oxidative stress. Oxidative stress is a condition in which free radicals are responsible for age related degenerative disease like Alzheimer's, Parkinson's and auto-immune disease. Sorghum has potential antioxidant competency which can act against reactive oxidative species. Pearl millet is most widely cultivated millet in eastern Uttar Pradesh. Consumption of pearl millet boosts hearth health, cure acidity problems, prevents gall bladder stones. It is also beneficial for asthmatic patients.

Table 1: Nutrition in Millets as per Compression of Rice and Wheat ¼100/grama½

Sr. No.	Carbohydrate (gm.)	Protein (gm.)	Fat (gm.)	Fiber (gm.)	Calorie (mg.)	Calcium (mg.)	Phosphorus (mg.)	Magnesium (mg.)	Zinc (mg.)	Iron (mg.)
Jowar	67.7	9.97	1.73	10.22	334	28	274	133	2.96	3.95
Bajara	61.8	10.96	5.43	11.49	348	27	289	124	2.76	6.42
Ragi	66.8	7.16	1.92	11.18	321	364	210	146	2.53	4.62
Kodo	66.2	8.92	2.55	6.39	332	15	101	122	1.65	2.34
Kutaki	65.6	10.13	3.89	7.72	346	16	130	91	1.82	1.26
Wheat	64.7	10.59	1.47	11.23	322	39	315	125	2-85	3.97
Rice	78.2	7.94	0.52	2.81	356	7	96	19	1.21	0.65

Source: FAO

Holistic Approach Towards Health

A holistic approach towards health considers entire person- mind, body and spirit as interconnected elements that contribute to overall well- being. It goes beyond the treatment of specific symptoms or diseases and takes into accounts various aspects of an individual's life and environment. Here are some key principles and components of a holistic approach towards health

i) **Mind-Body connection**: Recognising the interconnectedness of mental and physical health is fundamental. Mental well being can influence physical health and vice versa. Practices like mindfulness, meditation and yoga aim to foster a harmonious mind body relationship.

ii) **Physical Health**: This involves the absence of illness or diseases and presence of good physical fitness. Adequate nutrition, regular exercise and access to healthcare services contribute to good physical well- being.

iii) **Nutrition and Lifestyle**: understanding the importance of a balanced diet, regular exercise and healthy lifestyle choices are crucial. Nutrition plays a crucial role in supporting overall health and lifestyle factors such as sleep, stress management and physical activities.

iv) **Emotional well- being**: Addressing emotional health is essential for holistic approach. This involves managing stress, cultivating positive relationships, provided emotional support when needed and engaging in activities that promotes joy and fulfilment.

v) **Social support**: Recognising the impact of social relationship on health is important. Strong social connections can provide emotional support, reduce stress and contribute to a sense of belonging and purpose.

vi) **Spiritual well- being**: For some individuals, spiritual aspects play a significant role in their overall health. It is a dimension of overall well –being that extends beyond physical and mental aspects, encompassing a wide range of beliefs and practices.

Thus, we can say a holistic approach towards health involves considering the whole person physically, mentally, emotionally and spiritually and recognizing the interconnectedness of various aspects of life.

Role of millets in gaining holistic health

Millets have gained attention for their nutritional benefits and their role in promoting holistic health. Here are some aspects of the role of the millets in gaining holistic health-

i) **Nutrient Density**: Millets are rich in essential nutrients such as vitamins, minerals, fibre, and antioxidants. They provide a well-rounded nutritional profile, contributing to overall health and well-being.

ii) **Diverse Nutritional Content**: Different types of millets offer varying nutritional compositions. For example, they may contain significant amounts of protein, dietary fibre, vitamins (such as B group vitamins), and minerals (such as iron and magnesium). This diversity supports a balanced and varied diet.

iii) **Rich in antioxidants**: Millets contain various antioxidants that help combat oxidative stress in the body. Antioxidants play a role in preventing cell damage and inflammation, supporting the body's defence against chronic diseases.

iv) **Dietary Fibre**: Millets are an excellent source of dietary fibre, which is essential for digestive health. Adequate fibre intake promotes regular bowel movements, helps prevent constipation, and supports a healthy gut micro biome.

v) **Gluten intolerance**: Many types of millet are naturally gluten-free, making them suitable for individuals with gluten sensitivities or celiac disease. Incorporating gluten-free grains can contribute to digestive comfort and overall well-being for those with gluten-related concerns.

vi) **Blood Sugar regulation**: Millets have a lower glycemic index compared to some other grains which means they have a smaller impact on blood sugar levels. This characteristic is beneficial for individuals managing diabetes or those seeking to maintain stable energy levels.

vii) **Cardiovascular Health**: The fibre, antioxidants and other bioactive compounds found in millets may contribute to cardiovascular health. These components can help lower cholesterol levels, regulate blood pressure and reduce the risk of cardiovascular diseases.

viii) **Weight Management**: Millets are nutrient-dense and can contribute to a feeling of fullness due to their fibre and protein content. Including millets in meals may support weight management by promoting satiety and reducing overall calorie intake.

ix) **Sustainable Agriculture**: Millets are hardy crops that can thrive in diverse climates with minimal water requirements. Their cultivation is often more sustainable and environmentally friendly compared to some other grains contributing to holistic health by supporting a sustainable food system.

x) **Cultural and Culinary diversity**: Millets have been staples in the diets of many cultures for centuries. Incorporating a variety of foods including millets adds diversity to the diet, providing a range of flavours, textures, and nutrients.

To fully harness the holistic health benefits of millets, it is essential that incorporate them into a well- balanced and varied diet. It is must to keep in mind that millets are full of fibres so before using them it is necessary to keep the ratio and quantity in mind regarding which type of millet you are going to use. Additionally, individual diet needs and preferences should be taken into account and if there is any health issues it is advised that before including millets in the diet, consultation with doctor is must.

Conclusion

Millets are highly nutritious and staple food in many countries. The major reasons behind cultivation and utilization pattern are increasing demand of fine cereals and processing of millets. Tedious processing techniques decrease the utilization of millets. These can help to overcome life style diseases and metabolic problems. They are cheap source of good quality proteins, dietary fibres, minerals to obtain good health and physical development. Millets are boon to fight against malnutrition and food insecurity. Policy makers should take some collective measures to encourage farmers to produce millet cultivation at larger level. It is not only good for health point of view but it has great significance on per capita water availability. Small scale enterprises can be started at local level to promote millet based food items. Now days, a lot of millet based food products are filled in market. Biscuits and confectionary items, beverages, flour, porridge, weaning foods, puffs and pops, flakes, pasta, vermicelli, multigrain dosa and idli mix etc are available in the market.

References

Chandel, G., Meena, R. K. and Dubey, M., 2014. Nutritional properties of minor millets: neglected cereals with potential to combat malnutrition, *Current Sci., 107(7), 1109 – 1111.*

Devi, P. B., Vijayabharathi, R., Sathyabama, S., Malleshi, and Priyadarshini, V. B., 2014. Health benefits of finger millet (Eleusine coracana L.) polyphenols and dietary fibre: a review. *Journal of Food Sci. Tech. (June 2014) 51 (6), 1021-1040.*

FAO, 1995. Sorghum and Millets in Human Nutrition, Food and Agriculture Organisations of United States, Rome, Italy.

Gopalan, C., Shastri, B., V. R., and Balasubramanian, S. C., 2014. Nutritive value of Indian foods, National Institute of Nutrition, Hyderabad.

17

Wheat: Challenges and Strategies for Sustainable Production in India

Himanshu Panday[1], Barsati Lal[2] Kamta Prasad[3] R.K. Rai[4] and Navneet Maurya[5]

[1,3,4]ICAR-Indian Institute of Sugarcane Research, Lucknow, Uttar Pradesh
[2]Department of Agricultural Economics, CoA, BUAT, Banda, Uttar Pradesh
[5]Department of Agricultural Extension, CoA, BUAT, Banda, Uttar Pradesh

Abstract

Agriculture is playing a crucial role in Indian economy, which contributes about 16 per cent share in total Gross Domestic Product (GDP) and total workforce engaged in agricultural and allied activities is 54.60 per cent. The total geographical area of the country is 328.70 million hectare. Out of which net cultivable land is 139.4 million hectare and gross cropped area is 200.20 million hectares with 143.6 per cent of cropping intensity. Wheat is the most important source of vegetative protein in the human diet, with higher protein content than other major cereals such as maize and rice. The total area of wheat in the world is 219 million hectares with a production of 760.92 million tonnes. The global average productivity is 3474 kg/ha. The production of wheat in China is 134.25 million tonnes, followed by India, Russia, USA, Canada, France, Pakistan, Ukraine, Germany and Australia with a production of 107.59, 85.89, 49.69, 35.18, 30.14, 25.24, 24.91, 22.17 and 14.48 million tonnes, respectively. In India the wheat crop is mainly grown in North India. The major wheat producing states are Uttar Pradesh, Punjab, Haryana, Madhya Pradesh, Rajasthan, Bihar, Gujarat, Maharashtra, Uttarakhand, West Bengal, Himachal Pradesh, Karnataka and Jammu & Kashmir. All these states contribute about 99.5 per cent of total wheat production in the country. Wheat productivity faces multiple challenges around the world, particularly the climate change. To achieve the target set for wheat production with respect to trans-challenges like burgeoning population, shift in consumption pattern, threat of climatic vulnerability, reduction in farmland size coupled with degradation of soil quality and demand for diverse products by the consumers is a major challenge. Sustainable food production is a must for achieving the food and nutritional security. Now Indian Farming production needs to be shifted to quality linked parameters for nutritional security.

Keywords: *Workforce, Challenges, Burgeoning Population, Sustainable Food Production and Nutritional Security etc.*

Introduction

Botanical Classification of Wheat

Kingdom :	Plantae
Phylum :	Tracheophyta
Class :	Liliopsida
Order :	Poales
Family :	Poaceae
Genus :	*Triticum*
Species :	*aestivum*

Botanical Description

Botanically wheat (*Triticum aestivum)* is a grass, widely cultivated for its seed. Commercially it is classified as a cereal grain that is a worldwide staple food. The many species of wheat together make up the genus *Triticum.* The most widely grown wheat are the hexaploid bread wheat (*T. aestivum*) and tetraploid durum wheat (*T. turgidum*). Other types of wheat, such as einkorn, emmer and spelt, are still grown on a more limited scale in some regions, though interest has been increasing of late. The archaeological record suggests that wheat was first cultivated in the regions of the Fertile Crescent around 9600 BC. Wheat plant has two types of root system, one is called seminal root system while other is clonal root system. The entire roots of wheat plant are adventitious and then plant has a permanent root system. After the 28-30 days of seedling emergence, seminal roots dry. The stem of wheat plant is erect, cylindrical and placed near to the ground known as tiller or primary tiller. Leave of the wheat plant have two parts: one is called sheath and other is blade. The inflorescence of wheat plant is known as spike or it is also called as ear. Each spikelet contains stamen, pistil and ovary. The wheat seed is called as caryopsis (the seed coat fused with pericarp is known as caryopsis).

Origin Place

Wheat originated in the Middle East, near the junction of national borders for the USSR, Iraq, Turkey and Iran.

Global Scenario of Wheat

It is the world most widely cultivated staple food crop, having been grown since prehistoric times and being consumed in various forms by over one thousand million people worldwide. It has been dubbed the "King of Cereals". It is widely used for human consumption in the form of flour, *suzi*, flour, maida, and is eaten by a variety of consumers in various ways such as *chapati, puries, paratha, dalia, halwa and upma*. Wheat has recently been used in processed food products such as baked leavened breads, biscuits, cakes, pastries, flakes and noodles. Wheat straw and wheat bran are also good sources of animal feed. Wheat is the most important source of vegetative protein in the human diet, with higher protein content than other major cereals such as maize (corn) and rice. It has a good nutritional profile with 12.1% protein, 1.8% lipids, 1.8 % ash, 2% reducing sugar, 6.7% pentose and 59 % starch. It is also a good source of vitamins, minerals and nicotinic acid.

The wheat is produced in about 120 countries of the world. The main wheat producing countries are China, India, USA, Russian Federation, Australia and Canada. The total area of wheat in the world is 219 million hectares with a production of 760.92 million tonnes. The global average productivity is 3474 kg/ha. The production of wheat in China is 134.25 million tonnes, followed by India, Russia, USA, Canada, France, Pakistan, Ukraine, Germany and Australia with a production of 107.59, 85.89, 49.69, 35.18, 30.14, 25.24, 24.91, 22.17 and 14.48 million tonnes, respectively (Ramadas *et al.*, 2020).

Table 1: Top five Wheat Producing Country in the World

Country	Area (M. ha)	Production (MT)	Yield (kg/ha)
China	23.38	134.25	5741.7
India	31.35	107.59	3431.1
Russia	28.86	85.89	2975.9
U.S.A.	14.87	49.69	3341.5
Canada	10.01	35.18	3512

Source: FAOSTAT (2020)

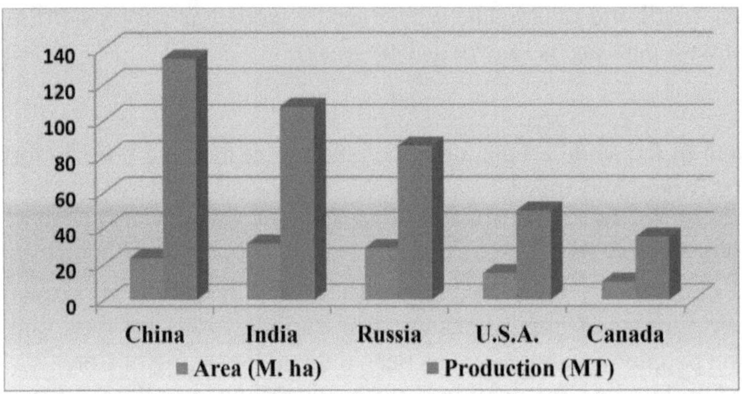

Indian Scenario of Wheat

Agriculture is playing a crucial role in Indian economy, which contributes about 16 per cent share in total Gross Domestic Product (GDP) and total workforce engaged in agricultural and allied activities is 54.60 per cent (*Census* 2011). The total geographical area of the country is 328.70 million hectare. Out of which net cultivable land is 139.4 million hectare and gross cropped area is 200.20 million hectares with 143.6 per cent of cropping intensity. At present the net sown area is 139.90 and net irrigated area is 68.6 million hectares in country. *(Annual Report* 2020-21, *DAC&FW*). Among the many crops grown in India wheat has special status as it is the staple food for about one-third of the population and a major supplement in the diet. It is the second most important crop next to rice in terms of area, production and contribution to the gross agricultural income of the country. In terms of yield per hectare, it stands first among all the food grains including rice. In fact, increase in the food grains production in India over the years has been attributed mainly to spectacular performance of the wheat crop. The year 1967-68 ushered in the Green Revolution. Introduction of high yielding varieties stepped up the pace of agricultural growth.

In India the wheat crop is mainly grown in North India. The major wheat producing states are Uttar Pradesh, Punjab, Haryana, Madhya Pradesh, Rajasthan, Bihar, Gujarat, Maharashtra, Uttarakhand, West Bengal, Himachal Pradesh, Karnataka and Jammu & Kashmir. All these states contribute about 99.5 per cent of total wheat production in the country. Remaining states like Jharkhand, Assam, Chhattisgarh, Delhi and other North Eastern States contribute only about 0.5 per cent of the total wheat production in the country. The cultivated area under wheat is 31.45 million hectare. Out of which, the Uttar Pradesh has largest share in area with (30.19 %) followed by Madhya Pradesh, Punjab, Haryana and Rajasthan 20.83%, 11.15%, 9.91% & 8.06%, respectively.

Table 2: Top five Wheat Producing States in India

State	Area (Million ha)	Production (MT)	Yield (Kg/ha)
Uttar Pradesh	9.50	32.59	3432
Madhya Pradesh	6.55	19.61	2993
Punjab	3.51	17.57	5008
Haryana	2.53	11.88	4687
Rajasthan	3.12	10.92	3501
All India	31.45	107.59	3421

Source: Directorate of Economics & Statistics, DAC&FW (2020-21)

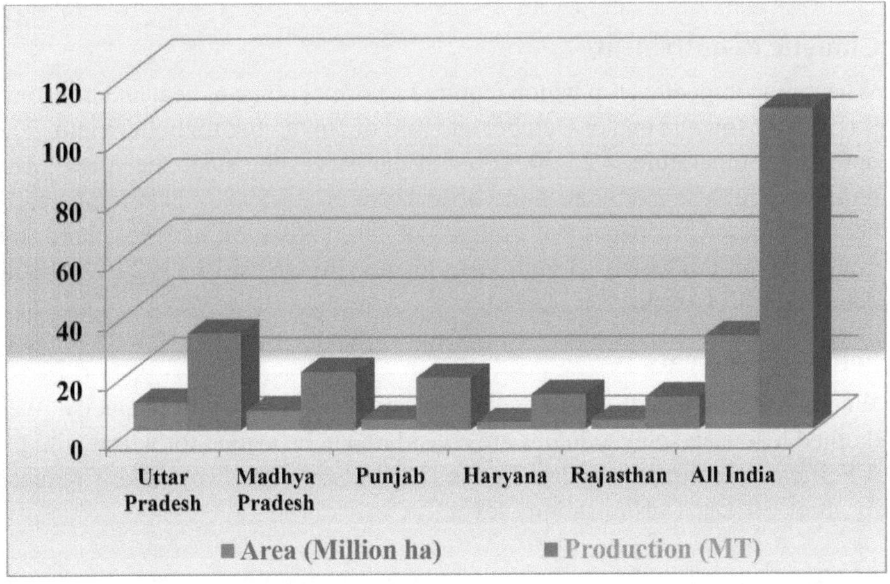

Wheat growing Mega Zones in India

Zone	Area
North Western Plaines Zone (NWPZ)	Punjab, Haryana, Delhi, Rajasthan (except Kota and Udaipur division) and Western UP (except Jhansi division), parts of J&K (Jammu and Kathua Distt.) and parts of HP (Una Distt. and Paonta Valley) and Uttarakhand (Tarai Region)
North Eastern Plaines Zone (NEPZ)	Eastern Uttar Pradesh, Bihar, Jharkhand, Odisha, West Bengal, Assam and Plains of North Eastern states
Central Zone (CZ)	Madhya Pradesh, Chhattisgarh, Gujarat, Rajasthan (Kota and Udaipur division) and Uttar Pradesh (Jhansi division)
Peninsular Zone (PZ)	Maharashtra, Karnataka, Andhra Pradesh, Telangana, Goa and Tamil Nadu
Northern Hills Zone (NHZ)	Western Himalyan regions of J&K (except Jammu and Kathua Distt.), Himachal Pradesh (except Una and Paonta Valley), Uttarakhand (except Tarai area); Sikkim and hills of West Bengal and N.E. States

(Source: Gupta *et al.,* 2023)

Cultivation Practices

Wheat, in India, is best grown as a rabi or winter season crop since the conditions during that time are conducive for growth and ensures maximum yield.

Climatic Requirement

Wheat is a temperate crop which required a normal range of heat and moisture. It is mostly sown in end of October and mid of November for better yield. The optimum temperature is 25-30 ^{0}C but at the harvesting stage the wheat plant requires a high temperature that varies between 35-50 ^{0}C. This temperature helps the plant in ripening of grains. For the cultivation of wheat crop, the minimum rainfall which is required during the growing season is 20”. It should not exceed 40 (Ahsan *et al.,* 2016).

Soil Requirements

Wheat needs soil with a moderate amount of water holding capacity. Well drained loams and clayey loams are considered to be a good for wheat with pH 6-8. It can tolerate salinity level up to (EC) 6ds/m, however, yield is reduced by about 50% at 14ds/m salinity level.

Zone Wise Recommended Varieties

A. Northern Hill Zone (NHZ)		
Early Varieties	**Mid Varieties**	**Late Varieties**
HS-277, Pusa Kiran (HS-542)	Pusa Saket (HS-507), HS-562,	Pusa Baker (HS-490), Shivalik (HS-420), HS-295, HS-207,
HPW-251	HPW-349, HPW-184	HD-2380
VL-616, VL-829	VL-738, VL-804, VL-907	
B. North Western Plains Zone (NWPZ)		
Karan Shivani (DBW-327), DBD-332	WB-2, PBW-723,	DBW-71, DBW-90, DBW-173,
Karan Vaidehi (DBD-370)	HPBW-01, HD-3086	WH-1021, WH-1124
WH-1270	DBW-88, WH-1105	PBW-590, HD-3059
C. North Eastern Plains Zone (NEPZ)		
Pusa Wheat (HD-3249)	K-1006, K-0307	Pusa Vatsala (HD-3118), Pusa Basant (HD-2985)
Karan Vandana (DBW-187)	HD-2733, Poorva (HD-2824), HD-2967, DBW-39, NW-5054	DBW-14, DBW-107, HI-1563, NW-1014
D. Central Zone (CZ)		
Karan Vashnavi (DBW-303), DBW-47, DBW-187	Mangla (HI-1077), Purna (HI-1544), GW-190, GW-273, GW-322, GW-366, Parvati (HD-2278), RAJ-2535	Kanishka (CG-1029), Pusa Ahilya (HI-1634)
E. Peninsular Zone (PZ)		
MACS-2971	HI-1633, HI-8802, HI-8805, HI-8826,	HD-3090, HD-2932
DDK-1025, DDK-1029	MACS-4058, MACS-4100, DDW-48	PBW-533, HUW-510, DWR-195
F. Southern Hills (Nilgiri and Palni Hills): HW-741, HW-1085, HW-2044		

Source: (Gupta *et al.*, 2018 and Gupta *et al.*, 2023)

Sowing Time

The normal time of sowing of high-yielding cultivars in irrigated areas start in the beginning of November. Long and medium duration cultivars like UP2003 and Arjun, etc. should be sown in the first fortnight of November. Short duration cultivars (120-125 days) like UP115, 162, etc in second fortnight of November. Under specific circumstances, wheat is also sown in December. In late-sown wheat, only short duration cultivars should be sown because there is comparatively less reduction in their yields as compared to long and medium duration cultivars. When wheat is sown beyond December there is a drastic reduction in yield. After November, delay in sowing by each day causes reduction of 5kg/ha/day in northeastern parts of the country and 41kg/ha/day in north-western and central parts of the country.

Seed Rate: Seed rate of wheat under various sowing methods varies in case of line sowing/kera/pora methods seed rate is 80-100 kg/ha Sowing by dibbling requires 25-30 kg seed per ha and in case of broadcasting method rate of seed is 100-120 kg/ha. In late sown condition seed rate is 125-155 kg/ha.

Spacing: For timely sown irrigated crop row to row spacing is 22-22.5 cm (Row) for late-sown irrigated conditions it is 15-18 cm.

Depth: For dwarf wheat under irrigated conditions, the planting depth should be between 5 and 6 cm, however, under unirrigated condition sowing depth varies from 8-10 cm.

Seed Treatment: Wheat seed is usually to be treated with vitavax or thyram @ 2.5 gm/kg seed. The seed of loose smut-susceptible varieties should be given solar or hot-water treatment.

Method of Sowing: Methods of wheat sowing are elaborated as below:

Broadcasting: This is an old method wherein seeds are broadcasted and then worked in by harrowing to cover them. The seeds are not uniformly distributed in the field, hence this method should not be encouraged.

Behind Desi Plough: Previously this method was used by the most of the farmers but now a days due to use of tractors for cultivation this method is rarely used. In this method, the seeds are dropped by hand behind the plough in the furrows. The seeds are dropped at a depth of 5 to 6 centimeters.

Dibbling: It is done with the help of a small implement known as 'Dibbler'. It is an iron or wooden frame with pegs. The holes are made with the help of dibbler and then seeds are dropped in the holes and covered by soil. Under this method, a uniform distance between plant to plant as well as row to row can be maintained. This method is best suited for sowing of newly released varieties as very less quantity of seed is available but by dibbling more area can be covered with limited seed without impairing the productivity. However, it is a time-consuming process.

Zero Tillage Technique: It is a new method used in the Rice-Wheat cropping system. Zero or minimum tillage provides minimum disturbance of the soil by placing the seeds directly in furrows. Seeds are then covered with welldecomposed compost and rice stubbles left in the field. Some advantages of zerotillage is the reduction of about 30 per cent in the use of water, compared to conventional tillage, as well as an improvement in the physical properties of soil. In addition, by planting wheat in time, higher yields may be obtained. Land should be moist at the time of planting wheat under zero tillage, so that the seed drill can be operated under unploughed land after the rice harvest. Farm yard manure (FYM) should be spread in the field at 5 tonne per ha with rice stubbles left in the field after the paddy harvest. An improved seed cum fertilizer drill attached with a roller and tines is run by a power tiller to plant five to six rows of wheat at a time. Seeds and fertilizers are kept in advance in separate boxes which will be dropped through separate holes fitted with plastic pipes. The tines fitted in the seed cum fertilizer drill at a recommended row to row spacing. This will simply scrap the land making a 2 to 3 cm furrow, remove weeds, and drop the fertilizer and seeds simultaneously. The holes for dropping fertilizer and seeds are calibrated in advance based on their recommended rates. This causes minimal disturbance to the soil and surface residues in the soil with a small furrow opening for seed and fertilizer placement. Wheat planting is normally done in early November. Wheat seeds can be planted as a relay crop under this condition. The recommended seed rate is 120 kg per ha.

Drilling: The seed is sown using a seed drill or ferti-seed drill. They are either bullock or tractor-powered. Seeds are dropped at depth using a drill, resulting in uniform germination and a regular stand. Seed drills, also known as ferti-seed drills, are widely available in the market.

FIRB System: The furrow irrigated raised bed (FIRB) has been recently developed and is being promoted by the Rice-Wheat consortium of the CGIAR institute. It has two types of configuration;

i. FIRB 2 rows (two rows on each raised bed of 70 cm with top width of about 40 cm and inter-row spacing of 30 cm),

ii. FIRBS 3 rows (three rows on each bed with inter-row spacing of 15 cm

In determining the dimensions of the beds, factors such as spacing between tractor tyres, soil types, rainfall and groundwater conditions, salinity, irrigation water quality and requirements of crops grown in rotation are of prime importance. Change over from growing crops in flat to ridge-furrow system of planting crops on raised bed alters the crop geometry and land configuration, offers more effective control over irrigation and drainage as well as their impacts on transport and transformations of nutrients. In furrow irrigated raised bed (FIRB) system, water moves horizontally from the furrows into the beds (subbing) and is pulled upwards in the bed towards the soil surface by capillarity, evaporation and transpiration, and downwards largely by gravity. Management of irrigation water is simpler and more efficient. On an average about 30% less irrigation water was required compared to flat bed method and improved crop yields by more than 20%. FIRB planting saved 30 to 50% wheat seed compared to flat bed planting. Better upland crop production is possible under wet spells because of proper drainage. Farmers can apply N and irrigation water at grain filling stage to improve protein content without inducing lodging. Reduced lodging can have a significant positive effect on yield as many farmers do not irrigate after heading precisely to avoid lodging. Weeds between the beds can be controlled mechanically early in the crop cycle. Herbicide dependence is reduced, and hand weeding and hoeing between rows are easier.

Nutrient Management

It is desirable that 10 tonnes of well decomposed manure or compost applied per hectare for wheat crop. The fertilizer requirement of the wheat crop is as follows:

For timely sown crop: 120Kg Nitrogen, 60kg Phosphorus and 40Kg Potash per ha.

For late sown crop: 80Kg Nitrogen, 40kg Phosphorus and 40Kg Potash per ha.

The 1/2 dose of nitrogen, full dose of Phosphorus and Potash should be applied at the time of sowing as basal dose. Remaining 1/4 nitrogen to be applied immediately after first irrigation and the other 1/4 nitrogen after second irrigation. Urea can also be applied as foliar spray at the late-tillering and late jointing stages if the crop shows nitrogen deficiency. If there is a deficiency of sulphur in the field, sulphur containing fertilizer such as ammonium sulphate or single super sulphate should be used. If green manure of dhaincha is used before the wheat crop, about 50-60 kg of nitrogen per hectare is saved.

Interculture

Usually no intercultural is given to the rainfed wheat crop but in case of irrigated crop one hoeing is done when the crop is about 18 days old to remove the weeds and prolongs the first irrigation. Generally weeding is done after 1 ½ to 2 months after sowing or weedicides like 2, 4 D, Avadex or Nitrofen are applied for controlling *Chenopodium sp, Angallis sp. Asphodelus sp. Phalaris sp.* of weeds.

Water Management

In wheat cultivation, irrigation requirement depends on various factors *viz.* type of soil, variety grown, area, temperature, etc.

Critical Stages of Wheat for Irrigation

Sl. No	Name of Critical Stage	No. of days after sowing
1.	Crown Root Initiation	21
2.	Tillering Stage	45
3.	Jointing Stage	65
4.	Flowering/Booting Stage	85
5	Dough Stage	120

Crop Rotation

Maize-Wheat 1 year, Maize-Wheat-Cotton-Barseem 2 years, Paddy-Wheat 1 year, Cotton-Wheat-Green gram 1 year, Paddy-Toria-Sugarcane-Ratoon-Wheat 3 years, Groundnut +Mung-Wheat 1 years and Early Paddy-Early Potato-Late Wheat 1 year.

Plant Protection

Diseases: Wheat crops suffer from several diseases causing reduced yield and quality. The major diseases and their chemical control are given blow:

Name of Disease	Causal organism	Chemical Control
Black or stem rust, Yellow or strip rust and Brown or leaf rust	Puccinia graminis *f. sp.* Tritici, Puccinia striiformis f. *sp.* tritici and *Puccinia* triticina (*P.* recondita)	Spray 0.2% zineb or Dithane M-45.
Alternaria leaf blight	*Alternaria triticina*	Treat seed with vitavax @2.5g/kg of seed, Spray 0.2% zineb or dithane M-45.
Powdery mildew	*Erysiphe graminis*	Spray propiconazole (Tilt 25EC) @ 0.1% at emergence or appearance of disease
Loos smut	*Ustilago tritici*	Seed treatment with *Trichoderma viride* @ 1.5g/kg seed
Karnal bunt	*Tilletia indica*	Seed treatment with ceresin G.N. @ 2.5g/kg of seed is beneficial

Insect-Pests

Wheat is attacked by a number of insects-pests and rodents both in the fields and in storage. Some insets-pests and their control are given as follows:

Name of Insect-Pests	Chemical Control
Termites	Treat the seed with chlopyriphos (1g/kg seed) or fipronil (0.3g/kg seed), broadcast chlopyriphos mixed soil (3lt. in 40-45 kg soil) 15 days after sowing.
Army worm and gram caterpillar	Spray of carbaryl (sevin 50 WP @ 2.5 kg/ha in 800lt. of water/ha.
Aphids	Spray of imidacloprid @ 20 g a. i. per ha

Harvesting

The time of harvesting of wheat depends on the type and variety grown. The wheat crop is usually harvested when the grains become hard and the leaves become dry and brittle. General time for wheat harvesting in different zones is given as under

Wheat Growing Zones	General Time of Harvesting
NEP zone	Last week of March and continues till mid April
NWP zone	Second fortnight of April
Central zone	End of February to March
Peninsular zone	Second fortnight of February to beginning of March
Hilly zone	May-June

Yield

40-45 q grain and 70-80 q straw/ha may be obtained from dwarf wheat varieties under irrigated areas. Under rainfed condition, 20-30 q grain and 60-70 q straw/ha may be obtained.

Challenges of Wheat Production

Wheat productivity faces multiple challenges around the world, particularly the climate change. Addressing these challenges requires continued innovation and collaboration across the international wheat research and breeding network. The network is global and diverse and covers all continents. (Langridge *et al.*, 2022)

1. **Increasing Population Pressure:** The increasing population exerts the utmost pressure on the existing wheat production. The global population is projected to reach 9.8 billion by 2050. Population growth and changes in consumption patterns, including new dietary preferences require the production of more (and more diverse) food. The average calorie intake has increased as populations have become richer. Excessive consumption and post-harvest losses and waste draw down scarce resources and increase environmental footprints, including the degradation of water quality. The need to produce more food implies an increase in land clearing for food production and the productivity of agricultural lands (Sagasta *et al.,* 2017). India ranks first in the list of countries by population. The current population of India is 143.90 corers based on the worldometer elaboration of the latest United Nations data. In India, the yearly per capita wheat consumption will increase to 74 kg in 2030 and 94 kg in 2050 from 60.4 kg in 2019. In case of India, under both low and high-fertility rate assumptions, aggregate wheat demand in India

will increase by 32-38% in 2030 and by 70-104% in 2050 compared to 2019 level of consumption (Mottaleb *et al.,* 2023). Increasing demand of food grains especially wheat will further put pressure on the fragile ecosystem that may aid in climatic abruptions.

2. **Environmental Pollution:** Both agricultural intensification and agricultural expansion have potential negative environmental impacts. The cultivation in degraded lands further damaged the soil structure and reduced water retention capacity. The consequences of greenhouse gas emissions on climate change are in part attributable to general farming practices rather than individual farmer land use decisions. The consequences of soil erosion and nutrient mining, on the other hand, can be vividly observed at the farm level. Globally, 24% of land experienced degradation from 1995- 2008 as measured by declining ecosystem function and productivity, particularly along sloped lands in Africa and Southeast Asia. In India, decades of twice-annual harvests through an intensive rice-wheat system have led to chronic land degradation and stagnating yields. The combined depletion of nitrogen, phosphorous and potassium in thirteen provinces in India to be as much as 80kg/ha annually, while other research suggests yield barriers in part driven by micro-nutrient deficiencies including zinc, boron, annual depletion of nitrogen, phosphorous and potassium, etc.

3. **Climate Change:** Climate change would force changes in diets around the world as some of the crops would be difficult to produce in the altered environment which can be subsidized by switching to crops that can thrive in those altered climates. Important crops like wheat and maize are more prone to be affected as they produce less grain at temperatures above 30°C. The impacts of climate change on food systems are expected to be widespread, complex, geographically and temporally variable, and profoundly influenced by socioeconomic conditions. The Asia-Pacific region is likely to face the worst impacts on cereal crop yields. Loss in yields of wheat, rice, and maize are estimated in the vicinity of 50, 17, and 6 per cent, respectively by 2050. This yield loss will threaten the food security of at least 1.6 billion people in South Asia. The projected rise in temperature of 0.5°C to 1.2°C will be the major cause of grain yield reduction in most areas of South Asia. In India, the physical impact of climate change would cause an increase in the average surface temperature by 2-4°C, changes in rainfall during both *kharif* and *rabi* months, a decrease in the number of rainy days by more than 15 days, an increase in the intensity of rain by 1- 4 mm/day and an increase in the

frequency and intensity of cyclonic storms (Sharma *et. al.,* 2015). India being a sub-tropical country with an agrarian economy is highly prone to the impact of climate change. Variability in climate is one of the biggest environmental threats to Indian agriculture, potentially impacting wheat production and food security.

3.1. Increase in Temperature: The Northern Indian states such as Uttar Pradesh, Punjab, Haryana, Uttarakhand, and Himachal Pradesh are some of the major wheat-producing states where the crop is more vulnerable at a 1°C rise in temperature resulting in reduction of wheat yield. In Haryana, night temperatures during February and March in 2003-04 were recorded at 3°C above normal, and subsequently, wheat production declined.

3.2. Drought and Heat stress: Drought and heat stress are becoming increasingly prevalent. Around half of all wheat producing area globally experiences periods of heat stress, and 20 million hectares or more routinely experience water deficits.

4. **Salt Affected Soil and Water Pollution:** Post-Green Revolution, wheat productivity has shown a remarkable increase but with increase in the use of fertilizers and plant protection chemicals has led to declining soil health. On the other hand, an increase in the area under irrigation resulted in a declining groundwater table. In India, about 4.5 million hectares salt-affected areas are under wheat cultivation posing a major problem for canal-irrigated areas. Even though soil amendments and proper drainage are the more constructive solutions, the pace of reclamation is not substantial. Water pollution is a global challenge that has increased in both developed and developing countries, undermining economic growth as well as the physical and environmental health of billions of people. The global growth of crop production has been achieved mainly through the intensive use of inputs such as pesticides and chemical fertilizers. The trend has been amplified by the expansion of agricultural land, with irrigation playing a strategic role in improving productivity and rural livelihoods while also transferring agricultural pollutants to water bodies. Water pollution from agriculture has direct negative impacts on human health; for example, the well-known blue-baby syndrome in which high levels of nitrates in water can cause met hemoglobin anemia – a potentially fatal illness – in infants. Pesticide accumulation in water and the food chain, with demonstrated ill effects on humans, led to the widespread banning of certain broad-spectrum and persistent pesticides (such as DDT and many organophosphates), but some such pesticides

are still used in poorer countries, causing acute and likely chronic health effects. Aquatic ecosystems are also affected by agricultural pollution; for example, eutrophification caused by the accumulation of nutrients in lakes and coastal waters has impacts on biodiversity and fisheries. Water-quality degradation may also have severe direct impacts on productive activities, including agriculture. For example, dam siltation caused by the mobilization of sediment due to erosion has cost many millions of dollars. Irrigation using saline or brackish water has limited agricultural production in hundreds of thousands of hectares worldwide. According to the Organization for Economic Co-operation and Development, developing countries alone, the environmental and social costs of water pollution caused by agriculture probably exceed billions of dollars annually. Quality water availability for irrigation is the single biggest factor influencing wheat yield. The irrigation with polluted water will significantly reduce the wheat yield (Sagasta *et al.,* 2017).

5. **Formation of Hard Pan in Soil Plough Layer:** The R–W systems are characterized by a diverse edaphic environment. Wet tillage (puddling; tillage in standing water) a pre-requisite under conventional tillage, requires repeated tilling in standing water to crush soil clods, eliminate macro-pores, reduce puddled layer strength and disperse fine clay particles. Repeated wet tillage of heavy to medium textured soils which is aimed at reducing percolation losses can cause compaction of soil below 7–10 cm of the soil surface. Wet tillage induces the formation of hardpan with higher soil bulk density that adversely affects root proliferation in terms of root geometry and architecture of succeeding wheat crop resulting in yield reduction.

6. **Green House Gases Emissions:** The greenhouse gas emission was significantly higher for R–W (by 0.2 kg C_{O2e} k^{g-1} grain) and maize–wheat (by 0.1 kg C_{O2e} k^{g-1} grain), compared with cotton–wheat cropping system. In contrast to rice, wheat requires an aerobic root zone and favors dry soil. Thus, the alternate aerobic and anaerobic regime affects the physicochemical and biological soil properties and alters the nutrient availability, root proliferation, moisture availability and finally the crop–root interactions.

7. **Declining Groundwater Tables:** The R–W system has aggravated the problems associated with deterioration of soil structure and declining groundwater tables resulting in decreased crop, land and water productivity. The conventional crop establishment techniques for the R–W system are highly exhaustive in terms of labor, water and power,

but especially water owing to the significantly higher water demand of rice and wheat crop. The annual per capita water availability in India is expected to decrease from 1600 m^3 to 1000 m^3 by 2025. It would also reduce the share of agriculture sector by 8–10% up to 2025 because of rising water demand by other allied sectors. The declining groundwater tables and over-fertilization are responsible for increased energy consumption and deteriorated water quality. Recently, NASA gravity mapping satellite 'GRACE' reported a drawdown of groundwater by 30 cm $year^{-1}$ from 440,000 km^2 area of NW India, indicating a decline of 0.04 m groundwater $year^{-1}$.

8. **Fragmentation of Land:** Over the years, a visible declining trend in farm holding size has been observed and is another major concern for the nation as a whole. This is caused by the fragmentation of farmland owing to the increasing population, nuclear families and the decline in cultivable areas due to urbanization. The average size of operational holdings has decreased from 2.28 hectares in 1970-71 to 1.84 hectares in 1980-81, to 1.41 hectares in 1995-96 and to 1.08 hectares in 2015-16 *(PIB, 2020)*. Considering the declining trends observed in the size of agriculture holdings in the past and the prospective increase in population over time, the fragmentation of holdings is likely to continue and the average size of operational holdings is expected to further decrease in the country. Holdings beyond certain lower size become less profitable due to absence of scale of economy. Further decrease in holding sizes poses major challenges for increasing the productivity of wheat as well as other crops.

9. **Unavailability of Improved Seed:** Adoption system and germplasm dissemination in India have been made in formal and informal ways. Even though new improved varieties are developed and made available to farmers by NARS but around 80% of all seeds are saved by the farmers. Seed replacement is most important for improving the productivity but it is very slow in wheat crop which poses a threat for development of the crop. A majority of farmers in India are small & marginal lack awareness of improved wheat varieties due to weak linkages, however, the development and adoption of improved varieties are crucial for achieving the targeted production of wheat.

10. **The Emergence of Newer Pests and Diseases:** As the year passes, the pests of wheat have developed some resistance even though controlled under contingent situations. A new range of pests and diseases have been

emerging due to changing climatic factors putting a serious constraint on wheat productivity.

11. **Declining Factor Productivity:** A major concern among policymakers is the declining total factor productivity over the years owing to stagnating yield levels with increased use of inputs and resource services. It is a major concern in the intensive cropping areas wherein rice and wheat are widely under cultivation. This can be countered by the adoption of improved technologies coupled with the use of optimal resources.

Future Strategies of Sustainable Wheat Production

To achieve the target set for wheat production with respect to trans-challenges like burgeoning population, shift in consumption pattern, threat of climatic vulnerability, reduction in farmland size coupled with degradation of soil quality and demand for diverse products by the consumers is a major challenge. Sustainable food production is a must for achieving the food and nutritional security. Now Indian Farming production needs to be shifted to quality linked parameters for nutritional security. If so, what kind of strategies should we adopted to make wheat production sustainable for the future population? However, there cannot be a single solution for this trans-problem and hence a concerted efforts to be made (Yadav *et al.,* 2010).

1. **Use of the innovative Technologies by the Farmers:** As India strives to meet the growing demand for wheat amidst mounting environmental and economic challenges, sustainable agricultural practices and technological innovations hold the key to the sector's future resilience. Adoption of climate-smart farming techniques, such as conservation agriculture, precision farming and agro forestry, can improve resource-use efficiency, soil health and resilience to climate change. The available technological interventions need to be evaluated and promoted to achieve higher land and water productivity, improve soil health, mitigate GHGs emissions and lessen water and C footprints. The technologies such as SMP-based irrigation scheduling which are cost and resource efficient and eco-friendly can help increase productivity, profitability and sustainability of the R–W system. There is a need to ensure large-scale initiatives that integrate awareness, technical know-how, on-farm demonstrations, subsidy on agri-machinery for crop residue management and financial incentives to farmers who retain crop residues on their fields. Breeding of climate resilient varieties having drought and heat stress tolerance must be in priority as climate change is inevitable. Varieties for problematic areas like saline and alkaline as

well as rainfed condition should also be developed. The promotion of short-duration wheat varieties helps in water saving due to their shorter life cycle, but adoption by the farmers was found to be less due to their lower yields. Some new short-duration wheat cultivars with higher yield potential should be bred and recommended to the farmers for adoption. The extension efforts need to strengthen to disseminate and popularize the RCTs among farmers and encourage them for their widespread adoption. The agricultural extension agencies could play a key role in educating the farmers with the conduct of front-line demonstrations and on-farm trials. Furthermore, investment in research and development for developing climate-resilient wheat varieties, coupled with capacity building and extension services for farmers, can foster innovation and knowledge dissemination across the agricultural value chain (Hussain *et al.,* 2020).

2. **Inclusion of Legumes in Rotation and Crop Diversification:** The continuous R–W cropping with intensified tillage in NW IGPs has shown negative impacts on land degradation and rapidly declining groundwater resources. Therefore, crop diversification with oilseed and pulses instead of growing rice had shown potential for reduced tillage, soil health improvement, reduced water requirements and increased water economy. Restoration of soil fertility and improvement in soil physicochemical and biological properties helped in achieving a higher yield of the next crop. Therefore, practicing the crop rotation with legumes between the two crops helps in enhancing soil N economy and contributes toward cropping systems' sustainability.
3. **Carbon Sequestration:** The R–W system is highly intensive in terms of C inputs and outputs through the addition of above and below-ground biomass and burning of rice and wheat residues. The burning of residues although seems an easy way out for the farmers because of large straw load and shorter window period between paddy harvesting and wheat sowing, yet residue burning has not been a viable option, as it leaves significant C footprints which negatively impact the C sustainability of R–W system. Among different residue management strategies, the development of Happy Seeder a modified zero tillage machine, capable of drilling wheat seed into the standing rice stubble without any preparatory tillage appeared highly cost-effective, eco-friendly and sustainable technology.
4. **Conservation Agriculture:** In South Asia, conservation agriculture (CA) was introduced for achieving the long-term sustainability of

agricultural production systems in the context of changing climate. Broadly, CA has been advocated with three basic principles, *viz.* diverse crop rotation, reduced tillage and mulching or residue retention. The resiliency of CA to climate change may be attributed to reduced labour & diesel costs, inclusion of legumes in the system and reducing soil moisture loss. The synergistic interaction of CA with improved R–W management practices such as timely sowing and efficient weed control measures certainly improves grain yield. Soil and water conservation measures to be make legally binding to avoid miss management of these resources. *In-situ* water conservation like check dam, bunding of farm fields, optimum tillage, innovative agronomic practices for weed management without use of chemicals, mulching, integrated farming system and use of Artificial Intelligence to be promoted. Clean and green technology for wheat production to be developed for arresting the degradation of production resources, pollution and enhancing the sustainability of agriculture as a whole.

5. **Optimisation of Nutrients use Efficiency:** Policies for nutrients use efficiency optimization to be initiated and popularized. Nitrogenous fertilizers are the major factors in water pollution due to faster leaching, hence slow release and leaching resistance fertilizers can be developed. These fertilizers should be popularized among the farmers to make the cropping sustainable and to save the precious water resources from pollution. Further, cost effective easily adoptable technologies for purification of polluted water to be developed. Proper use of purified water should also be ensured.

6. **Water-Saving Technologies:** The real water-saving technologies are those which minimize the unproductive evaporation losses and partition the share of unproductive evaporation to transpiration to favor inflow of soil nutrients to the plants through roots. Among different short-duration cultivars and optimum date of planting are the real water-saving technologies advocated for the farmers. These technologies aim at reducing the water losses that cannot be economically recaptured; however, their efficiency varies spatially and temporally. These technologies could be used to reduce the undesirable drainage losses especially in regions where the water could be recaptured. For effective saving of irrigation water, irrigation water management for the entire sequence including the fallow period, instead of just for rice or wheat individually, should be considered.

7. **Government Initiatives and Support:** Recognizing the significance of wheat in ensuring food security and rural livelihoods, the Government of India has implemented various policies and schemes to support wheat farmers. Subsidized inputs, such as seeds, fertilizers, and irrigation equipment, aim to enhance farm productivity and income levels. Additionally, price support mechanisms, including MSPs and procurement operations by agencies like the FCI, provide price stability and market access to farmers, thereby incentivizing wheat cultivation (Singh *et al.,* 2008). Support is also provided to farmers (including small and marginal farmers) through initiatives and programmes of the Government like Interest Subvention Scheme, MIDH, NFSM, Neem Coated Urea, PMKSY, PMFBY and NMSA etc.

Conclusion

Wheat crop has been a part of human civilization and is regarded as one of the important food grains and crucial component in agri-food system of the global agriculture. It is the most widely grown crop in the world and provide a fifth of food calories and protein to the word population. The exploitation of natural resources and adverse impacts of climate change on crop productivity demanded research communities to explore the crop production and management measures. The demand for wheat is also rising at a higher rate than the population growth rate, particularly in Asia. With continued global population growth rate and growing popularity of wheat based processed food in the world there is a need for continued efforts to ensure further transformation of wheat agri-food system, including sustainable intensification of wheat production to stay within planetary boundaries. Further, changing climatic conditions pose a threat to the production and productivity of wheat which directly affect the food security. Despite India achieving the self-sufficiency status in wheat production but still there are many challenges which need to be resolved.

References

Anonymous (2019-20). Agricultural Statistics at a Glance. Ministry of Agriculture and Farmers' Welfare, GOI, New Delhi.

Anonymous (2020-21). Annual Report: Department of Agriculture, Cooperation and Farmers Welfare, Utter Pradesh., agricoop.nic,in

Anonymous (2020-21). Annual Report: Department of Agriculture and Statistics, Utter Pradesh. agristatics.nic. in

Anonymous (2021-22). UPDES. Production quantities of wheat by district. Retrieved, from http://updes.up.nic.in.

Bhatt, R., Singh, P., Hossain, A., & Timsina, J. (2021). Rice–wheat system in the northwest Indo-Gangetic plains of South Asia: issues and technological interventions for increasing productivity and sustainability. Paddy and Water Environment, 19(3), 345-365.

Gupta, A., Kumar V., Kumar P., Pal R., Gurudyal, Singh C., Tyagi B.S., and Singh G. (2023). Compendium of Wheat Varieties Notified in India during 2018-2023. Research Bulletin No. 51, ICAR- Indian Institute of Wheat & Barley Research, Karnal- 132001, India: pp 32

Gupta, A., Singh, C., Kumar V., Tyagi B.S., Tiwari V., Chatrath R., and Singh G.P. (2018). Wheat Varieties Notified in India since 1965. ICAR- Indian Institute of Wheat & Barley Research, Karnal- 132001, India:101 pp

Honoured, C. O. (2020). Press Information Bureau Government of India.

Hussain A., Iqbal A., Hayat Khan Z. (2020). Introductory Chapter: Recent Advances in Grain Crops Research. Recent Advances in Grain Crops Research. Intech-Open. Available at: http://dx.doi.org/10.5772/intechopen.90701.

Mateo-Sagasta, J., Zadeh, S. M., Turral, H., & Burke, J. (2017). Water pollution from agriculture: a global review. Executive summary, 35.

Mottaleb, K. A., Kruseman, G., Frija, A., Sonder, K., & Lopez-Ridaura, S. (2023). Projecting wheat demand in China and India for 2030 and 2050: Implications for food security. Frontiers in Nutrition, 9, 1077443.

Peter Langridge, Michael Alaux, Nuno Felipe Almeida, Karim Ammar, Michael Baum, Faouzi Bekkaoui, and Bettina Berger, (2022). "Rapid technological advances provide many additional opportunities and options for improving sustainable wheat production, but there are several constraints that may limit our ability to grasp and leverage these opportunities. The major constraints can be summarised under four categories: Meeting the Challenges Facing Wheat Production" The Strategic Research Agenda of the Global Wheat Initiative" Agronomy 2022, 12(11), 2767.

Ramadas, S., Kiran Kumar, T. M., & Pratap Singh, G. (2020). Wheat Production in India: Trends and Prospects. Intech Open. doi: 10.5772/intechopen.86341

Saljok, A., Hafiz, M. and Lissan, U. A. (2016). "Botanical description of Wheat and its production in Pakistan" ticles/major-crops/botanical-description-of-wheat-and-its-production-in-pakistan.

Sharma, I., Tyagi, B.S., Singh, G., Venkatesh, K. and Gupta, O.P., (2015). Enhancing wheat production-A global perspective. The Indian Journal of Agricultural Sciences, 85(1), pp.03-13.

Singh, A., Singh, S. and Singh, S. (2008) "Economic analysis of production constraints in wheat crop: case of North-Western India". Agricultural Situation in India. 65(3): 145-149.

Yadav, R.; Singh, S.S.; Jain, N. and Prabhu, K.V. (2010). Wheat Production in India: Technologies to Face Future Challenges, Journal of Agricultural Science ISSN: 1916-9752, Vol. 2, No. 2, June 2010 E-ISSN: 1916-9760-164.

18

Development of Climate Resilient Village: An Animal Husbandry Perspective

Diksha Patel[1] and Rahul Kumar Rai[2]

[1]Krishi Vigyan Kendra, Banda, Uttar Pradesh

[2]Department of Agricultural Economics, CoA, BUAT Banda, Uttar Pradesh

Abstract

The development of climate-resilient villages is crucial for ensuring sustainable livelihoods, particularly in the context of increasing climate variability and extreme weather events. This abstract focuses on the perspective of animal husbandry in fostering climate resilience within rural communities. Climate-resilient village development entails a multi-faceted approach that integrates traditional knowledge with modern techniques to mitigate the impacts of climate change on agriculture and livestock rearing. In the context of animal husbandry, this involves strategies such as improving livestock breeds that are better adapted to local climatic conditions, implementing water management practices to ensure adequate hydration for animals during droughts, and adopting agroforestry systems to provide shade and fodder. Furthermore, enhancing veterinary services and disease management practices are essential components of climate-resilient animal husbandry, as changing climate patterns can exacerbate the spread of diseases. Additionally, promoting diversified income sources through activities like apiculture and aquaculture can help buffer against income loss during climate-related shocks. Community participation and capacity building are fundamental for the success of climate-resilient village initiatives. Empowering local communities with knowledge and resources enables them to identify climate risks, implement adaptive measures, and strengthen their resilience to future climate challenges. In conclusion, integrating climate resilience into animal husbandry practices is imperative for sustainable rural development. By employing a holistic approach that combines traditional wisdom with innovative solutions, communities can build resilience and thrive in the face of climate conversions.

Keywords: *Climate, Animal Husbandry, Shade, Fodder, Veterinary Services, etc.*

Introduction

Changing climatic conditions are projected to affect food security from the local to global level. The predictability in rainy season patterns will be reduced, while the frequency and intensity of severe weather events such as floods, cyclones and hurricanes will increase; other predicted effects will include prolonged drought in some regions; and water shortages; and changes in the location and incidence of pest and disease outbreaks. Growing demand for biofuels from crops can place additional pressure on the natural resource base. New policy driven options are required to address the emerging challenges of attaining improved food security. Agricultural production and food security are strongly interlinked, and the impact of climate change would have major implications in determining socio economic fabric of all nations and the globe. It has been declared since long that climate changes have been happening, and the need for adaptations is stressed upon as an immediate response, although emphasis on mitigation of impacts is parallel. The challenges posed by climate change to agriculture and food security require scientific understanding of the impacts through global and local scale approaches leading to formulation and implementation of desired immediate and long term actions. The need for a climate resilient approach to agriculture is critical for India where more than 80 percent agriculturists are small-holder farmers. India recites in villages. India has total 640867 villages where 68.84% of Indian population are living and 65 % is dependent on agriculture & 70 m on livestock farming. India's share of world income was 22.6% in 1700, as against 3.8% in 1952 and 1.5% at 2014-15. Indian agriculture today faces a multipronged set of challenges pressured simultaneously by several sectoral and non-sectoral demands. All this is further aggravated by the extreme weather variations that are being experienced. The majority of farmers are small and marginal landowners who are resource-poor. They are most affected due to their low adaptive capacity and risk-taking ability. By incorporating various adaptation measures in the agriculture system one can increase the resilience and adaptive capacity of the small land holders.

Climate Change

The climate change is increasingly becoming an important consideration for our lives. It refers to the difference in the earth's global climate or in regional climate over a period of time. It reflects change in average weather conditions or in distribution of weather events over periods of time that ranges from decades to millions of years. It affects all the aspects of climate, making the rainfall less predictable, changing the seasons, and increasing the occurrence of extreme events such as heat waves, cold waves, drought and flood etc. These changes can be caused by the processes like External forcing mechanisms (solar output, orbital variation) and Internal forcing mechanisms (plate tectonics, volcanism, human activity etc.).

Climate change has been considered as one of the most serious long term challenges to be faced by the farmers. Especially in India being an agrarian economy where the small and marginal farmers are contributing the maximum share in milk production are belonging to resource poor category. Climate change impacts on livestock are being witnessed at around the world, but in developing country like India is much more vulnerable due to large section of the population depends on agriculture for their livelihood. The warming trend in India over the previous hundred years was estimated to be 0.60°C. The (IPCC, 2007) predicted that by the year 2100 increase in global average surface temperature may be between 1.8 and 4.0°C, with global average temperature increases by 1.5-2.5°C. Approximately 20-30 percent of plant and animal species are expected to be at risk of extinction (FAO, 2007). In developing countries climate change may affect the growth in milk and meat production. Thus the focus on food and nutritional security needed to be broadening to adopt a sustainable livelihood approach. Food production is identified to be subset of the wider livelihood consideration of rural people. Hence, a shift from the materialistic perspective on food production to a social perspective, with focus on the enhancement of people's capacities to secure their own livelihood was found inevitable.

Nature and Importance of Climate Change

According to Maplecroft climate change vulnerability index (2011) India is the world's most vulnerable country apart from Bangladesh. In India, 25 states are prone to various disasters except volcanic activity. Out of that flood and drought are most abundant and most frequent causing severe loss to economy every year. India faced many disasters due to climate change in last few years.

Year	Event
2002	All India drought Severe cold wave (2002-03)
2004	Drought like situation High temperature anomaly in March
2005	High temperature in January
2006	Floods in arid Rajasthan & AP Drought in high rainfall NE India
2007	High temperatures in Jan□Feb
2009	All India drought
2010	Warmest year
2011	Failure of Sep rains in AP
2012	Drought in Punjab, Haryana, Gujarat, Karnataka, Cyclone & Floods in AP
2013	Drought in Bihar & Jharkhand, Floods in Uttarakhand, Phailin cyclone
2014	Floods in J&K, Cyclone Hudhud, widespread hailstorm in March
2015	Chennai flood
2019	Cyclone Fani
2020	Cyclone Amphan
2022	Cyclone Mandous, Cyclone Sitrang
2023	Cyclone Biparjoy

Impact of Climate Change in Agriculture

- Cereal productivity to decrease by 10-40% by 2100.
- Greater loss expected in rabi. Every 1°C increase in temperature reduces wheat production by 4-5 million tons.
- Increased droughts and floods are likely to increase production variability
- Changing climatic conditions affect crop growth and livestock performance, availability of water, the functioning of ecosystem
- Extreme weather events are likely to become more severe and frequent
- Crop yields from rain-fed agriculture would be reduced and crop net revenues could fall by as much as 90% by 2100
- Yield reduction of 4.5-9% likely to impact 1.5 % of GDP
- The burning of stubble leads to the outflow of air pollution. Each year, during October and November; this extensive agricultural burning lasts for more than three week. Due to air pollution about 900 million people-one eighth of the world's population get affected

Impact of Climate Change on Animal Husbandry

1. Dry matter intake Decreases 10 to 20% in commercial dairy herds.
2. The length and intensity of the estrus period decreases.
3. Decreased conception (fertility) rate.
4. Decreased fetal growth and calf size.
5. Decreases in milk production can range from 10 to 25%.
6. Delayed puberty in bulls due to non-availability of balanced feeding during summer.
7. Reduction in bodyweight
8. Inability to move
9. Increased body temperature
10. Emergence of new livestock disease
11. Mortality and morbidity will increase

Climate Resilience: It is the capacity of socio-ecological system to absorb stress and maintain function in the face of external stress imposed upon it by climate change and to adopt, recognise and evolve into more desirable configuration that improve the sustainability of the system, leaving it better prepared for future climate change impacts.

Approaches to Cope up with Climate Change in Agricultural Sector

- A special package for adaptation should be developed for rain fed areas based on minimizing risk. The production model should be diversified to include crops, livestock, fisheries, poultry and agro forestry; homestead gardens supported by nurseries should be promoted to make up deficits in food and nutrition from climate related yield losses. Farm ponds, fertilizer trees and biogas plants must be promoted in all semi-arid rainfed areas which constitute 60 percent of our cultivated area.
- A knowledge-intensive, rather than input-intensive approach should be adopted to develop adaptation strategies. Traditional knowledge about the community's coping strategies should be documented and used in training programs to help find solutions to address the uncertainties of climate change, build resilience, adapt agriculture and reduce emissions.
- Conserving the genetic diversity of crops and animal breeds and its associated knowledge, in partnership with local communities must receive the highest priority.

- Breed improvement of indigenous cattle must be undertaken to improve their performance since they are much better adapted to adverse weather than high performance hybrids. Balancing feed mixtures, which research shows has the potential to increase milk yield and reduce methane emission, must be promoted widely.
- An early warning system should be put in place to monitor changes in pest and disease profile and predict new pest and disease outbreaks. The overall pest control strategy should be based on Integrated Pest Management because it takes care of multiple pests in a given climatic scenario.
- A national grid of grain storages, ranging from Pusa Bins and Grain Golas at the household/ community level to ultra- modern silos at the district level must be established to ensure local food security and stabilize prices.
- The agriculture credit and insurance systems must be made more comprehensive and responsive to the needs of small farmers. For instance, pigs are not covered by livestock insurance despite their potential for income enhancement of poor households. A special climate risk insurance should be launched for farmers

Climate Change Extension

The extension approach and services that specifically address ecological repercussion, over exploitation of natural resources, potential problems arising from global warming and climate change, bringing mass awareness about harmful effect of climate change, motivating people to adopt mitigation practices through demonstration of climate change technologies; identification, documentation and validation of Indigenous Traditional Knowledge (ITK) are called climate change extension. For this purpose extension agent should acquire comprehensive knowledge and understanding of the science of global warming and climate change so that extension programme can be oriented accordingly.

Extension Approaches

1) **Research and capacity building centre**: ICAR has divided country into 13 major and 127 micro level agro-climatic zones. According to Prof. M.S. Swaminathan, Research and capacity building centre are to be established in each of the 127 agro-climatic zones for enabling farm families to meet the challenges arising from abnormal monsoon behaviour. Each one of the centre should have a metrological station and

a weather information bureau.

2) **Development of good weather code**: For each of the agro-climatic zone drought, flood and good weather code should contain detailed for managing drought and flood. The good weather code will indicate the steps needed to maximize the benefits of a good monsoon and minimize the adverse impact of aberrant monsoon.

3) **Climate risk manager:** According to Prof. M.S. Swaminathan at least one woman and one male member of every block/ panchayat should be trained in as Climate risk manager in order to help the village to handle situation like drought flood and sea level rise in coastal region in an effective manner. Such climate risk manger should become well versed in the science and art of disaster management.

4) **Community based weather forecasting:** Traditionally farmer monitor rainfall and other weather parameter based on their indigenous knowledge on weather. These measurements usually gave an idea about rainfall pattern but farmers lacked exact quantification. Many farmers in semi-arid India has established their own weather observation systems, using manual rain gauge, thermometers and wind anemometers to plan their farming activities the localised . the local weather observations are interpreted by the farmers in combination with the localised with localised weather and climate forecasts to take up appropriate farm decisions.

5) **Climate field school:** Asian Disaster Preparedness Centre (ADPC), Bangkok has introduced the concept of climate field school, to improve the basic knowledge of farmers on climate forecast used in designing crop management strategies. The aim of field school is to familiarized participation to process of learning by practice. The scientists of KVKs and line departments have to play a major role in this direction for effective preparedness and reducing risk.

Conclusion

Climate change affects dairy sector both directly and indirectly, the direct form of climate change on dairy sector manifest in the form of decrease productive and reproductive performances of dairy animals and eventually Increase incidences of diseases, especially parasitic diseases. Besides the direct effects of climate changes on animal production and animal health, there are profound indirect effect as well, which include climatic influences on quantity and quality of feed and fodder resources, water resources, etc. so the farmers should aware about the harmful effect of climate change. Immediate steps are needed to help farm families to enable them against climate change, Resilience at farm level should be coordinated and planned to avoid possible internal incoherencies among resilience measures and inconsistencies of resilience with wider rural development objectives. Certain resilience measures are contradicting water conservation objectives and create internal incoherencies, Crop and livestock enterprises could buffer the some of the impacts to some extent. Government should make some approaches, methodologies and technologies for marching towards climate resilient with livestock enterprises.

References

Balachandran, C. 2022. Impact of Climate Change on Livestock Health and Production. In Impact of Climate Change on Livestock Health and Production (pp. 1-20). CRC Press.

Choudhary, B.B., Singh, P., Gururaj, M., Kumar, R., Sirohi, S. and Kumar, S. 2020. Climate change and livestock sector in India: issues and options. Climate change and Indian agriculture: challenges and adaptation strategies (Eds. Srinivasarao, Ch, T Srinivas, RVS Rao, N Srinivasarao, SS Vinayagam, P Krishnan). ICAR-NAARM, Hyderabad. 517-538.

Glantz, M.H. Gommes, R. and Ramasamy, S. 2009. Adaptation and Mitigation. in: Coping with a changing climate: considerations for adaptation and mitigation in agriculture. www.fao.org.

http://wotr.org/publications/towards-resilient-agriculture-changing-climate-scenario

http://www.imd.gov.in/

http://www.ncipm.org.in/nicra/

https://en.wikipedia.org/wiki/Climate_resilience

IPCC (2007). Climate Change 2007: Synthesis Report: Contribution of Working Groups I, II and III to the Fourth Assessment Report of the Intergovernmental Panel on Climate Change, Core Writing Team, Pachauri R K and Reisinger A , IPCC, Geneva, 2008: 45.

Mondal, S. 2014. Text book of Agricultural Extension with Global Innovations. Kalyani Publishers, India.

Nirmala, G. and Pankaj, P.K. 2021. Livestock and Climate Change-A Gender Perspective. Climate Resilient Animal Husbandry, 76.

Perez, C., Jones, E.M., Kristjanson, P., Cramer, L., Thornton, P.K., Förch, W. and Barahona, C.A. 2015. How resilient are farming households and communities to a changing climate in Africa? A gender-based perspective. Global Environmental Change, 34: 95-107.

Sahai, S. 2012. Coping with Climate Change. Agri Cultuers. 14(2): 20-22

Sikka, A., Prasad, Y.G. and Rao, C.S. 2015. Development of Climate Resilient Villages. Global Science conference. Indian Council of Agricultural Research, New Delhi, India.

Thomas, C.K. and Sastry, N.S.R. 2013. Livestock Production Management. Kalyani publishers. India.

Upadhyay, R.C., Hooda, O.K., Aggarwal, A., Singh, S.V., Chakravarty, R. and Sirohi, S. 2013. Indian livestock production has resilience for climate change. Training compendium on climate resilient livestock and production system, November, 18.

Colour Plates

Chapter 5: Marketing of Feed and Fodder in India: Issues, Challenges and Way Forward

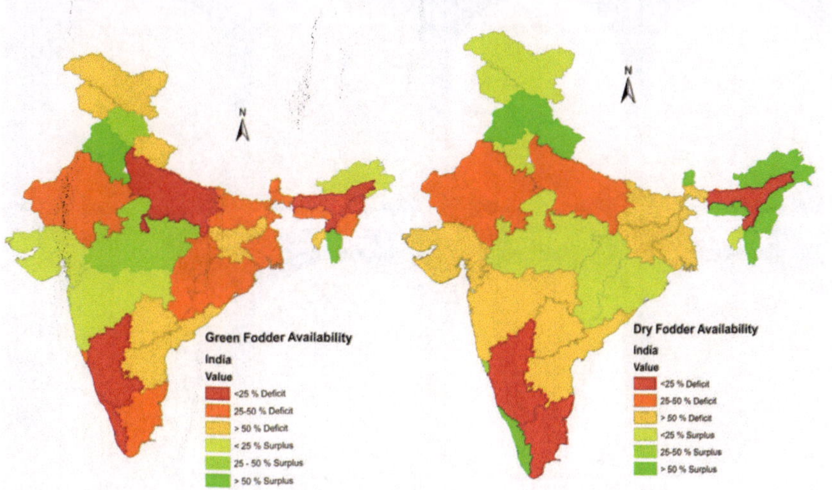

Source: Roy et al. (2019)

Fig. 1: Green and dry fodder availability scenario across Indian states

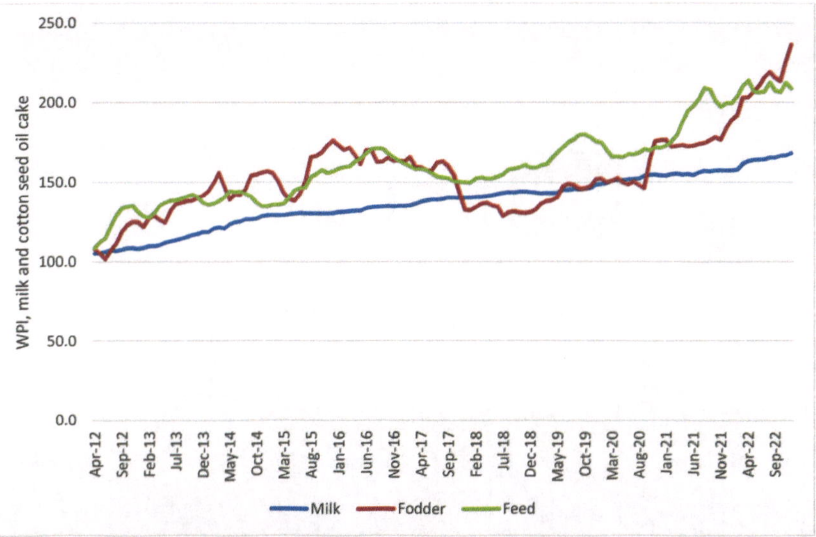

Source: Ministry of Commerce and Industry, GoI

Fig. 2: Wholesale price index of cattle feed, fodder and milk (Base 2011-12)

Chapter 10: Modern Prospective of the Ancient Grain: Millets

Chapter 11: Reform of Agricultural Market by e-NAM and It's Role in Agriculture Sector in Madhya Pradesh

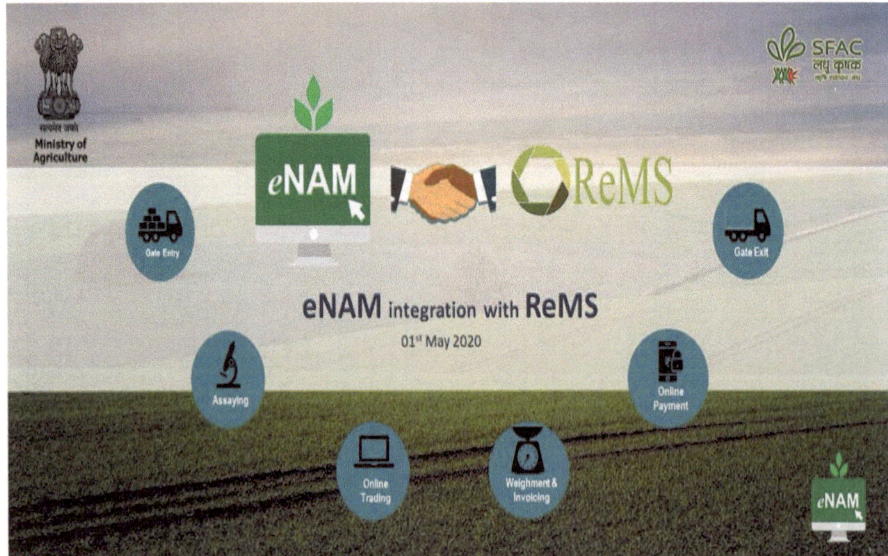

Fig. Shows the e NAM marketed process

Chapter 14: Recommended Doses of Fertilizers in Horticultural Crops

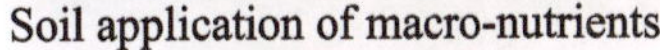

Soil application of macro-nutrients

Foliar application of micronutrients

Foliar application of Micronutrients

Chapter 17: Wheat: Challenges and Strategies for Sustainable Production in India

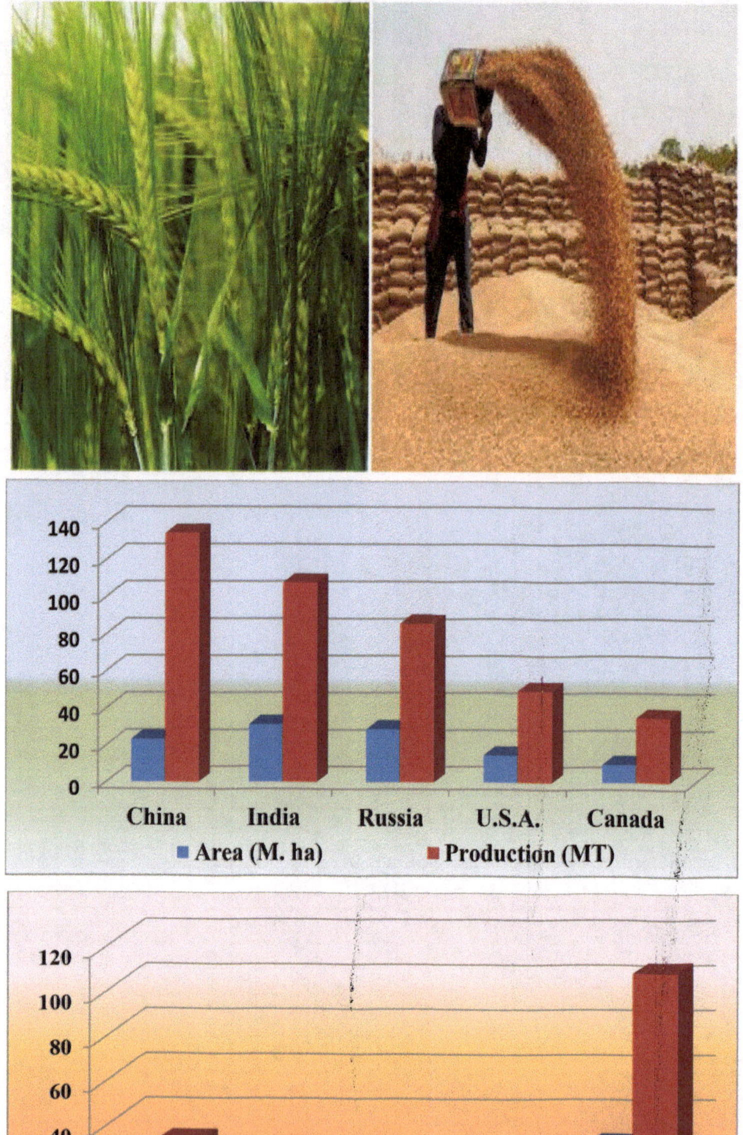

Index